DESIGN OF SMALL ELECTRICAL MACHINES

DESIGN AND MEASUREMENT IN ELECTRONIC ENGINEERING

Series Editors

D. V. Morgan,
School of Electrical, Electronic and Systems Engineering,
University of Wales College of Cardiff,
Cardiff, UK

H. L. Grubin,
Scientific Research Associates Inc,
Glastonbury, Connecticut, USA

THYRISTOR DESIGN AND REALIZATION
P. D. Taylor

ELECTRONICS OF MEASURING SYSTEMS
Tran Tien Lang

DESIGN AND REALIZATION OF BIPOLAR TRANSISTORS
Peter Ashburn

ELECTRICAL CHARACTERIZATION OF GaAs MATERIALS AND DEVICES
David C. Look

COMPUTERISED INSTRUMENTATION
Tran Tien Lang

PRINCIPLES AND TECHNOLOGY OF MODFETs VOLUMES 1 AND 2
H. Morkoç, H. Unlu, G. Ji

RELIABILITY OF GALLIUM ARSENIDE MMICs
Edited by Aris Christou

INTEGRATING QUALITY AND RELIABILITY INTO MICROELECTRONICS MANUFACTURING
Aris Christou

DESIGN OF SMALL ELECTRICAL MACHINES
E. S. Hamdi

DESIGN OF SMALL ELECTRICAL MACHINES

E. S. Hamdi
Dean of Engineering and Environmental Design
University of Wales, College of Cardiff, UK

JOHN WILEY & SONS
Chichester • New York • Brisbane • Toronto • Singapore

Baffins Lane, Chichester
West Sussex PO19 1UD, England
National Chichester (0243) 779777
International (+44) 243 779777

Other Wiley Editorial Offices

John Wiley & Sons, Inc., 605 Third Avenue,
New York, NY 10158-0012, USA

Jacaranda Wiley Ltd, 33 Park Road, Milton,
Queensland 4064, Australia

John Wiley & Sons (Canada) Ltd, 22 Worcester Road,
Rexdale, Ontario M9W 1L1, Canada

John Wiley & Sons (SEA) Pte Ltd, 37 Jalan Pemimpin #05-04,
Block B, Union Industrial Building, Singapore 2057

British Library Cataloguing in Publication Data

A catalogue record of this book is available from the British Library

ISBN 0 471 95202 8

Typeset in 10/12pt Times by Lasertext Ltd., Stretford, Manchester, UK
Printed and bound in Great Britain by Bookcraft (Bath) Ltd.

In memory of
my beloved son, Ahmed,
who suddenly departed while
this book was in production

CONTENTS

PREFACE

In the first half of this century there were three or four classics which covered the design of all types of conventional machines. Since the early fifties there have been no new texts dealing with the design of mainstream d.c. and induction motors. The need for a new book on machine design arises because of recent advances in applications and design techniques and the emergence of new materials; a 3.7 kW motor now weighs a lot less than one designed in the fifties.

This book covers the design of small and medium electrical machines and is written for practising engineers as well as postgraduate students who intend to specialise in machine design. Therefore a knowledge of electrical machine theory is assumed. The scope of the book is limited to small machines: the mechanical and thermal considerations associated with the design of large machines are quite different.

Electrical machine design is a multi-disciplinary subject. It involves not only electromagnetic design but also selection of engineering materials, thermal and mechanical analyses, materials and components specifications and production techniques. This book forms a self-contained and unified treatment of the subject of the design of small electrical machines. Background material and design data are given in the appendices. Throughout the text reference is made to relevant International and National Standards.

In the first two chapters the various engineering materials used in small machines, the types of enclosures and cooling arrangements, and the analysis and modelling of the heat transfer problems are introduced. Then, general concepts and constraints of machine design are discussed: the relationship between size, duty and rating, and between specific loadings and performance characteristics.

The book gives the constructional details of wound-field and permanent magnet d.c. motors, three-phase induction motors, brushless d.c. motors and line-start and variable-speed permanent magnet synchronous motors.

Design procedures are presented for wound-field and permanent magnet d.c. motors and three-phase induction motors. These are applied to produce typical designs for an industrial wound-field d.c. motor, a permanent magnet d.c. motor for automotive applications, a 50 Hz three-phase induction motor and a 400 Hz three-phase induction motor.

The application of the finite element method to the design of electric motors is described and an extensive review of the force and torque computation methods is presented.

I am grateful to my friend and colleague Derek M. German who read through the manuscript and made many useful suggestions, and to the publishers for inviting me to write the book. My thanks are due to Professors D. Vernon Morgan and Ronald T. Waters for their encouragement and support. Finally and most importantly, I would like to thank my wife, Soad, my daughter, Samar, and my sons, Ahmed and Omar, for giving me a great deal of moral support, and for putting up with my interminable sessions with the word processor.

Essam S. Hamdi
Cardiff, 1994

1

ELECTRICAL ENGINEERING MATERIALS

Materials used in an electric machine may be grouped into: (1) soft magnetic materials, (2) electrical conducting materials, (3) insulating materials and (4), in some cases, permanent magnets. Soft magnetic materials form the magnetic circuit in an electric machine. An ideal material for this purpose is one exhibiting the highest possible permeability and saturation flux density, and the lowest possible core loss. However, economic considerations, a phrase that is used frequently in this book, together with the machine's performance specifications, dictate the material to be used for a given application. For example, fractional-horsepower motors used in domestic appliances are, generally, optimised for minimum cost, and, therefore, they utilise relatively inexpensive, silicon-free, soft magnetic materials offering moderate characteristics. On the other hand, machines designed to operate at high frequencies necessitate the use of high performance, more expensive, materials in order to limit the machine's size and its losses to an acceptable level. Conducting materials, such as copper, aluminium and copper alloys, are used in electric machines to form low resistance electric circuits, while electrical carbon materials are used for making brushes.

Insulating materials are characterised by their high dielectric strength, high resistivity and high thermal conductivity. They are primarily used in electric machines to provide electrical separation between various current-carrying conductors at different voltages and between the slots housing them, without impeding the heat flow.

Permanent magnets are used in place of electromagnets in some electric machines to provide the excitation field. Permanent magnet materials may be grouped according to their basic chemical composition into Alnicos, ceramics (or ferrites) and rare earth magnets. The choice of a magnet material is mainly governed by the application requirements, which obviously include economic factors.

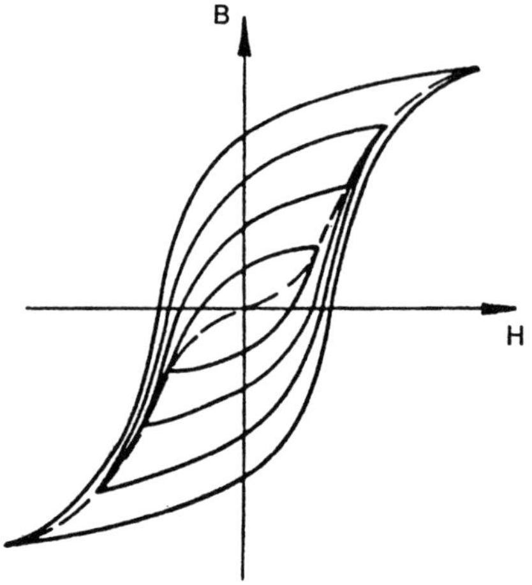

Figure 1.1 Family of hysteresis loops.

1.1 SOFT MAGNETIC MATERIALS

1.1.1 Iron losses

The iron losses or core losses produced in a magnetic material operating in an alternating magnetising field are traditionally separated into two components: hysteresis loss, and eddy current loss.

Hysteresis loss

Hysteresis loss is due to a form of intermolecular friction. The loss per cycle is proportional to the area enclosed by the hysteresis loop on the B–H characteristics of the material. The hysteresis loss increases as the maximum magnetic field reached during the cycle increases; as is illustrated in Figure 1.1.

Steinmetz [1.1] developed an empirical formula expressing the hysteresis loss (p_h, W/m^3) in terms of the maximum flux density (B, T) and frequency (f, Hz) as follows:

$$p_h = \eta B^n f \tag{1.1}$$

where η is a material constant and n is an exponent which has a value between 1.6 and 2. This loss is closely related to the coercivity, and processing of materials to reduce coercivity also reduces hysteresis loss.

Eddy currents and eddy current loss

The term "eddy current" refers to circulating electric currents which are induced in a sheet of a conducting material when it is subjected to alternating magnetic field. These eddy currents produce power which is dissipated as heat. The eddy current loss (p_e, W/m^3), at frequencies which are low enough for the inductive effects to be neglected, is given by the general equation [1.2]:

$$p_e = \frac{\pi^2 B^2 t^2 f^2}{\rho \beta} \tag{1.2}$$

where t is the thickness (m), ρ is the resistivity of the material (Ωm) and β is a coefficient which has different values for different geometries (equals 6 for thin laminations). Examining equation (1.2) shows that minimising the thickness of the material reduces the eddy current losses. It is economical, at 50 Hz, to go to the trouble and expense of dividing the iron used in electric machines into thin laminations (typically between 0.35–0.65 mm) and to insulate them from one another. At 400 Hz, thinner laminations made from special alloys or from grain-oriented electrical steels are used.

It may be noted that, in addition to the hysteresis and eddy current losses described above, another loss component exists; namely, rotational loss [1.3]. This is due the variation of the direction of the magnetic flux lines in the lamination plane. As yet, there is no established method for estimating this loss component.

1.1.2 Soft iron and silicon–iron

Soft magnetic materials employed in electric machines are, ideally, required to have high permeability (to reduce the reluctance of the magnetic circuit), high saturation flux density (to minimise the volume and weight of the iron parts) and to have low losses (as losses affect the efficiency and temperature rise). Practically, it is impossible to optimise all of these properties in a single material. This is because there are a large number of factors which affect magnetic properties (chemical composition, mechanical treatment and thermal treatment are the most important), and a conflict exists between obtaining low losses and high permeability as is illustrated by Figure 1.2 [1.4].

Ordinary soft iron contains impurities which adversely affects its characteristics. The most harmful impurity is carbon as it causes magnetic ageing (a slow deterioration of magnetic properties with time), and raises the coercive force and hysteresis loss. Carbon, and other harmful impurities, such as nitrogen and oxygen, can be removed by annealing for several hours in the presence of hydrogen.

The addition of a small percentage of silicon to soft iron improves several of its qualities: the resistivity increases, the coercive force considerably decreases and the stability of magnetic characteristics over time improves. Iron alloyed with 3% silicon has four times the resistivity of pure iron [1.5]. However, the presence of silicon results in a marginal decrease of the saturation flux density. Additionally, adding more than about 5% silicon to the iron makes it brittle and extremely difficult to work with. This restricts the maximum silicon content to 3.4% in commercial material (and a minimum of about 2.2% exists for grain oriented-steel below which proper orientation can not be achieved [1.6]). It may be noted that silicon is an inexpensive element, but silicon–iron materials are more expensive than soft iron owing to the mechanical and thermal processing of the alloy.

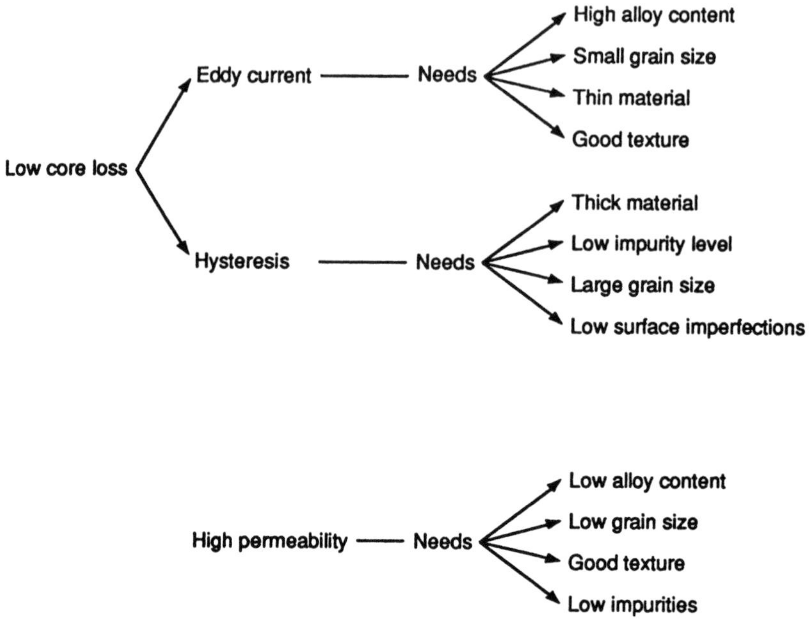

Figure 1.2 Conflicting requirements for obtaining low losses and high permeability in silicon–iron [1.4].

1.1.3 Electrical steels

Electrical steels (better known to the designers of electric machines as lamination materials) may be treated as two types; namely, non-oriented and grain-oriented steels.

Containing between 0 and 3% silicon, non-oriented electrical steels are essentially isotropic. They are typically produced in strips up to 1.3 m wide and between 0.35 and 0.8 mm thick. The sheet is usually supplied covered with a thin insulating layer to provide interlaminar insulation. Steam-blueing of one surface is usually used to insulate laminations used in small motors, while organic or inorganic coating is usually applied to one or both surfaces of large laminations. Insulation on both sides has the advantage of protecting the laminations from corrosion during storage.

Non-oriented laminations may be supplied fully annealed, ready for punching and core fabrication, or in a semi-processed state. When in the semi-processed (partly decarburised) state, the material is semi-hard, offering the advantage of better punchability, which yields longer tool life. However, a semi-processed material requires a final stress-relaxation and decarburising anneal at between 800 °C and 840 °C (depending on the chemical composition) to fully develop its magnetic properties [1.5].

Non-oriented semi-processed laminations have standard thicknesses of 0.5 mm and 0.65 mm, with typical specific loss values (W/kg), at 1.5 T, 50 Hz, that are in the range 4.55–5.5 and 6.0–7.1, respectively. Standard thicknesses of the fully processed material are 0.35 mm, 0.5 mm and 0.65 mm, and the guaranteed specific loss values (W/kg), at 1.5 T, 50 Hz, are 3.0, 4.0 and 8.0, respectively.

Table 1.1 Properties of Armco's grain-oriented steels

Trade name	Thickness (mm)	Typical core loss at:	
		1.5 T, 50 Hz	1.7 T, 50 Hz
Armco oriented	0.18	0.70	—
	0.23	0.77	—
	0.27	0.85	1.24
	0.35	1.11	1.57
Tran-CorH	0.23	—	1.01
	0.27	—	1.11

Table 1.2 Electrical steels standards

Country	Non-oriented	Grain-oriented
International	Euronorm 106 (1984)	Euronorm 107 (1986)
International	IEC 404-8-4 (1986)	IEC 404-8-5 (1987)
France	NF C 28-900 (1983)	NF C 28-920 (1975)
Germany	DIN 46 400, Pt.1 (1983)	DIN 46 400, Pt.3 (1975)
Japan	JIS C2552 (1978)	JIS C2553 (1975)
UK	BS 6404, Pt.8, 8.2–8.4 (1986)	BS 6404, Pt.8, 8.7 (1988)
USA	ASTM, A677, A677M (1984)	ASTM, A665, A665M, A725, A725M (1984)

The first commercial grain-oriented silicon–iron strip was produced in 1939 by Armco (Armco Electrical Steel, Pennsylvania, USA.). It was 0.32 mm thick, with a specific loss around 1.5 W/kg, at 1.5 T, 50 Hz. Although the production process of these materials has been significantly simplified over the years, it is still more elaborate than the production method of the non-oriented material. This additional processing increases the cost of the material, and these materials are not normally used in standard industrial motors.

Grain-oriented laminations are supplied fully processed, in widths typically from 20 mm to over 900 mm. They have standard thicknesses of 0.27 mm, 0.3 mm and 0.35 mm. However, thinner materials commercially exist, but they are not yet included in all the standards. Table 1.1 gives the thickness and specific loss for various grades of grain-oriented laminations produced by Armco. It should be noted that comparable grades are offered by various other manufacturers worldwide.

1.1.4 Specifications of electrical steel grades

The guaranteed maximum losses of the various grades of electrical steels described above are covered by various standards. Unfortunately, the permeability is not specified as a guaranteed value. Table 1.2 summarises the most widely used national and international standards.

In the *IEC Publication 404-8*, each material is identified by two hyphenated numbers and a letter followed by a number. The first number represents 100 times the guaranteed loss (W/kg), and the second number is 100 times the lamination thickness in millimetres. The letter identifies the grade. Letters A, D and E denote non-oriented fully-processed material, non-oriented semi-processed material and non-oriented alloyed (has a silicon content) sheet delivered in the semi-processed state, respectively. Grain-oriented materials are designated by the letters N, S and P, for a normal loss material, a reduced loss material and a high permeability material, respectively. The number which follows the letter representing the material code is one tenth of the frequency (Hz) at which the loss figure is specified. With the exception of the reduced loss and the high permeability materials, where the guaranteed losses are specified at 1.7 T, the guaranteed loss values are specified at 1.5 T.

A material identified as 250-35-A5 is a non-oriented steel, delivered fully annealed (A), 0.35 mm thick (35), whose guaranteed loss at 1.5 T, 50 Hz (5), is 2.5 W/kg (250). Similarly, the material 089-27-N5 is grain-oriented, 0.27 mm thick with guaranteed losses of 0.89 W/kg at 1.5 T, 50 Hz (this is the new code for the M4 grade). In most cases steel products have not changed much; only some adjustments of grading limits are made in the interest of standardisation. It, undoubtedly, will be some time before designers become fully familiar with the new codes.

It should be noted that electrical steels are tested and graded under pure sinusoidal conditions. These ideal conditions hardly represent the magnetising conditions in an electric machine. The presence of flux harmonics increases the core losses of electrical steels, and an allowance must be made for this at the design stage. This is of particular importance in motors designed to operate with inverters.

1.1.5 Special purpose alloys and high frequency laminations

Iron–cobalt alloys (Permendur) offer the highest saturation flux density, typically 2.35 T. With regard to their use in electric machines, two grades are identified: 24 Permendur and 49 Permendur. Containing 24% cobalt, the 24 Permendur alloy can be cast, which permits forming of parts of complex shapes such as magnetic pole pieces. With 49% cobalt and 2% vanadium, hot and cold–rolling of the 49 Permendur becomes possible, and the mechanical properties improve. Such an alloy, in spite of its cost, finds important applications in aircraft generators and some 400 Hz motors, where the high value of saturation flux density makes it possible to save volume and weight. As 24 Permendur alloy contains only 24% cobalt, it is cheaper than the 49 Permendur. However, the relative permeability of the 49 Permendur is significantly higher. Table 1.3 gives some of the typical properties of the Permendur alloys [1.7].

Iron–nickel alloys (permalloy), containing between 30% and 80% nickel and a small amount of other metals, are commercially produced. Of all the commercial materials, these alloys have the highest permeability and the lowest coercivity. On the other hand, their saturation flux density is very low; typically 0.7 T (36% Ni) to 1.3 T (78% Ni). Because of this, nickel–iron alloys do not find many applications in electrical machines. Some a.c. two-phase servomotors, primarily designed for

Table 1.3 Typical properties of Permendur alloys [1.6]

	49 Permendur	24 Permendur
Saturation induction (T)	2.35	2.35
Maximum relative permeability	7000	2000
Resistivity ($\mu\Omega$m)	0.40	0.20
Density (kg/m^3)	8070	7905

defence applications and operate at 400 Hz, utilise iron–nickel laminations. In such applications, the advantages of low loss and high corrosion resistance outweighs the disadvantages of high cost and low saturation levels.

Specifications for thin magnetic steel strips for use at frequencies equal to or above 100 Hz are given in the International Standards *IEC 404-8-8* (1991). The standard thicknesses are 0.05 mm, 0.1 mm, 0.15 mm for both oriented and non-oriented strips; and there is an additional standard gauge of 0.2 mm for non-oriented strips. These thin strips are designated by two letters, NO and GO for non-oriented and grain oriented, respectively, followed by a number representing 100 times the thickness. For example, NO 5 designates a non-oriented 0.05 mm strip. The production process and chemical composition thin magnetic steel strips are not covered by standard specifications. The specified guaranteed maximum specific loss at 1 T, 400 Hz, is between 13 and 15 W/kg for non-oriented strips. The guaranteed maximum specific losses for GO 10 and GO 15 at 1.5 T, 400 Hz, are 15 and 16 W/kg, respectively.

1.1.6 Amorphous metals

Produced by rapid cooling (cooling rate of about one million degrees per second) of alloys consisting of iron (about 80%), nickel and/or cobalt together with one or more of the following metalloids (a metalloid is an element or compound exhibiting both metallic and non-metallic properties): boron, silicon and carbon, amorphous metals have very soft magnetic properties [1.8]. Their core losses are typically 30% of those for grain-oriented electrical steels at 1.5 T, 50 Hz [1.8]. Their coercivities are an order of magnitude smaller than, and their permeabilities are about an order of magnitude greater than, the respective values for electrical steels. However, amorphous metals have several disadvantages when considered for application in electrical machines. Their saturation flux density is low, typically 1.5 T. Additionally, they are produced in thin ribbons of between 0.025 mm and 0.04 mm. This is an inherent limitation introduced in the manufacturing process in order to achieve the required quenching rate. The thin gauge and the roughness of the ribbon's surface limit the stacking factor achievable to less than 0.75. Another obstacle to the use of amorphous metals as a core material in electrical machines is their extreme hardness, typically over C-80 Rockwell, which renders standard cutting and manufacturing methods prohibitively costly.

Several techniques of producing a slotted, helical-wound, amorphous metal core were investigated by the General Electric Company [1.9], but these were either

Table 1.4 Test data comparison of two 370 W motors [1.10]

	Electrical steel core	Amorphous core
Stack length (mm)	44.4	88.9
Core weight (kg)	3.0	3.8
Full load efficiency (%)	65	78
Quarter load efficiency (%)	38	64

found to produce cores of unacceptable qualities or to be too slow to be acceptable for production use. It should be noted that, as far as the laboratory technique is concerned, Johnson [1.9] has succeeded in producing a slotted, helical, amorphous core. Also, he identified areas where developments are needed in order for his method to become commercially acceptable.

It may be useful, at this stage, to try to quantify the potential benefits obtainable by using amorphous metals for electric machines core materials. Obviously, a significant reduction in core losses can be achieved. Core losses comprise between 30% and 40% of the total full load losses in single-phase motors, and typically 15% in polyphase machines. This suggests that there is more to gain from using amorphous cores in single-phase motors.

In reference [1.10] a comparison is given between the performances of two 370 W, 60 Hz, 4-pole, single-phase induction motors; one using a conventional electrical steel core and the other employing an amorphous core. The stator core of the latter was built from several thousand amorphous laminations. Because of the large number of amorphous laminations needed to make such a small motor, a stacking factor of 0.67 was obtained, compared to 0.93 for the conventional motor. The lower saturation flux density of the amorphous material and the lower stacking factor resulted in a longer stack. A photo-etching process was utilised to make the lamination [1.10]; a process which is too slow to be considered for volume production. The rotor utilised an electrical steel core and an aluminium cage winding. Table 1.4 gives test data comparison of the amorphous core motor and the conventional motor [1.10].

There is no doubt that significant reduction in core losses, and increase in efficiency, can be realised in amorphous core electric machines. However, at present, the disadvantages of the amorphous metals are more significant. There is a better market for amorphous materials in wound core distribution transformers, and in some specialised low current applications where they can compete with iron–nickel alloys.

1.1.7 Selecting core materials

The objective in designing and manufacturing electric machines is to produce, as economically as possible, a machine which fulfils certain specifications and guarantees. Thus, a good design can only be evolved when economic and manufacturing aspects are given the same importance as technical considerations. Some examples of how the selection of core materials is influenced by the application requirements are now discussed.

Domestic appliances fractional-horsepower motors: The efficiency of these single-phase motors is typically about 60–70%. Low cost and suitability for volume production are of prime importance. Low material cost and ease of stamping point in the direction of a non-oriented, non-alloyed (silicon-free), semi-processed material; such as 890-50-D5.

Standard three-phase induction motors: This is by far the most widely used motor in industrial applications. The efficiency of a 250 W motor is typically about 70%, rising to over 90% for a 18.6 kW motor. Economic considerations are, of course, still important, but a balanced design, generally, necessitates the use of fully annealed, non-oriented, low loss electrical steels with core losses of about 4 W/kg.

The use of grain-oriented electrical steels and special purpose alloys is limited to 400 Hz machines. Grain-oriented electrical steels are also used in some turbo-generators and large high efficiency machines, but these are not covered in this book.

Stepper motors and d.c. servomotors are usually designed with a high flux density (1.8 T), and efficiency is rarely considered as an initial design parameter. Normally, they use thin gauge non-oriented electrical steels, but some designs utilise grain-oriented grades stacked in a random angular manner.

Solid core materials are normally used in parts of magnetic circuits carrying steady flux; such as field frames of d.c. machines. They are required to have high permeability particularly at high values of flux density. The principal materials used are: cast iron (used in field frames when cost is of primary importance and extra weight can be tolerated), purified iron and silicon–iron (d.c. electromagnets pole shoes), soft steel (frames) and ferro-cobalt (pole pieces).

1.1.8 Core fabrication

With few exceptions, most machine manufacturers do not deal directly with steel manufacturers. They obtain ready-stamped laminations from laminations suppliers who offer a comprehensive range of standard laminations. Figure 1.3 shows a stator and rotor lamination for a standard induction motor.

Small patches of laminations outside the standard stamped range, such as those required for R&D work, can be made more economically using special techniques; such as single slot punching (single notching or progression tooling) and laser notching. One supplier identifies a threshold of 300,000 laminations below which production from progression tools is cheaper than with single slot tools for sizes up to 340 mm in diameter [1.11]. Progression tooling yields better precision; as identical slots are produced by the same tool.

The action of the punch and die causes burrs to be formed on the underside of the lamination and, with pre-insulated laminations, the size of the burrs must be

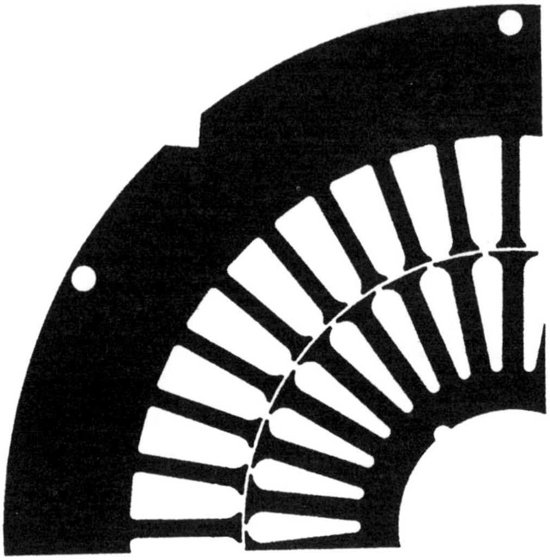

Figure 1.3 Small induction motor stator and rotor lamination.

Table 1.5 Properties of laminations insulating coatings [1.11]

	Phenolic resin	Synthetic resin	Phosphate
Coating	1 or 2-sided	2-sided	2-sided
Coating thickness (μm)	2–8	1–2	1
Resistance (1-sided) (Ωcm^2)	50	>90	10
Heat resistance (continuous)	180 °C	180 °C	850 °C
Corrosion resistance	Very good	Good	Good
Oil resistance	Good	Good	Good
Resistance to freon	Good	Very good	Good
Moisture absorption	None	None	None
Weldability	Low	Good	Good

kept to a value less than the thickness of the insulation film to prevent short-circuiting laminations in the core. Sharp tools minimise the burr size, and laminations are usually inspected after a set number of operations for excessive burr size. Properties of four grades of lamination insulating materials are given in Table 1.5 [1.11].

The laminations are, in most cases, assembled into a core pack, and lining bars located in slots are usually used to ensure correct alignment of the stack. The complete stack is then pressed and the laminations are welded together on the core back. As far as small machines are concerned, the weld is usually applied directly to the laminations in specially punched grooves. On large machines, core bars are welded to the lamination back to provide additional mechanical support. Finally, any high spots on the laminations are removed with a file. This, however, must be minimised and done with care to ensure that the laminations are not short-circuited.

1.2 CONDUCTING MATERIALS

These materials are obviously required to have the highest possible conductivity and the lowest possible temperature coefficient of resistance. They are also required to have adequate mechanical strength to enable fabrication of wires, coils and other intricate sections (such as commutator segments). Although conductors used in small electric machines are not normally subjected to high mechanical stresses, the process of making coils and inserting them in slots necessitates certain mechanical properties. Additionally, good weldability and solderability ensure high reliability and low electrical resistance of the joints.

1.2.1 Copper and its alloys

Copper is a highly tractable and ductile metal; it can be cast, forged, rolled, drawn and machined. Because of its high electrical conductivity and excellent mechanical properties, copper is the most widely used electrical conductor for the winding of electric machines.

Round copper wires coated with an insulating varnish are universally used for windings of small electric machines. Rectangular conductors are not normally used in low voltage (up to 500 V) small machines.

Copper is also used to make commutator segments for d.c. and a.c. commutator motors. This is mainly because the required stringent manufacturing tolerances can be achieved with copper, for it has good contact behaviour with carbon brushes and because small patches of copper segments can be manufactured economically. The latter is important as the number of commutator segments vary from one machine to another; depending on the design of the armature winding. Of course, the high conductivity of copper is still an advantage, but it is less important. Sometimes alloys with higher resistivity and better machinability than copper are used.

1.2.2 Copper alloys

The electrical and thermal conductivities of pure copper are only exceeded by silver. Therefore, alloying copper with other metals, except silver, results in lower conductivity.

Adding small percentages of tin, cadmium, beryllium and other metals to copper results in alloys with improved mechanical properties, but with higher resistance than that of pure copper. Adding 1 to 2.5% of beryllium to copper produces an alloy with improved corrosion resistance which can be formed into springs and strips. The resistivity of such an alloy is between three to six times that of copper, and it is used to make current-carrying springs and brush-holders.

Cadmium–copper alloys, containing 1.1% cadmium, are stiffer, harder, and have higher tensile strength than hard-drawn copper. The resistivity is marginally higher than that of copper, but the alloy can be flame-brazed. Cadmium copper is

Table 1.6 Properties of copper and aluminium

	Copper	Aluminium
Specific weight (kg/m^3)	8.89	2.70
Melting point (°C)	1083	658
Resistivity (Ωm)	1.17×10^{-8}	2.87×10^{-8}
Temp. coefficient of resistance at 20 °C	0.0039	0.004
Coefficient of linear expansion at 20 °C	17×10^{-3}	23×10^{-3}
Specific heat (cal/g °C)	0.093	0.022

sometimes used for making commutator segments and cage bars.

There are several other copper alloys, such as copper–silver and brass (66% copper and 34% zinc), but these are not normally used in small electric machines.

1.2.3 Aluminium

Aluminium is lighter and cheaper than copper and has a lower melting temperature (easier to cast). On the other hand, the conductivity of pure aluminium is only about 60% that of copper and it has low mechanical strength, and as such, it can not be drawn into fine wires. Table 1.6 provides comparison between the physical properties of copper and aluminium.

The principal use of aluminium in electric machines is in manufacturing cage windings for small induction motors. Utilising aluminium–silicon alloys containing between 6 to 12% silicon, the cage winding may be fabricated by diecasting; a simpler process than brazing cage bars to end rings. It may be noted that diecasting of a copper winding is also possible, but it requires higher casting temperature.

1.2.4 Wire gauge

Conventionally, copper and aluminium wires are specified in the United States by the American (or Brown & Sharpe) Wire Gauge (AWG), and in Britain by the Standard Wire Gauge (SWG). However, there is now an international wire specification (IEC 182-1). The wire tables conforming to IEC 182-1 are included in Appendix 1. As many designers are still more familiar with the conventional AWG and SWG wire gauge systems, a conversion table is provided in Appendix 1. It may be noted that in addition to the standard range a greater selection (known as split gauge numbers) is available from various suppliers.

1.3 INSULATION MATERIALS

Electric machines are energy converting devices, and during the process of energy conversion some loss of energy is inevitable. Losses occur, primarily, in the active parts of the machine; iron or core losses in the magnetic circuits and copper losses in the electric circuits. The energy lost within a machine is converted into heat which causes the temperature of the iron and copper parts to rise. Therefore, the losses in an electric machine are of importance not only because they affect its efficiency, but also because they can cause the windings temperature to rise significantly.

The temperature of the various parts of a machine is determined by the rates of heat production and dissipation, and a steady state temperature rise is reached when the two rates are equal. The active materials used in an electric machine (iron and copper) can withstand much higher temperatures than the limiting values for most insulating materials. Although reducing the specific electric and magnetic loadings reduces the losses and the heat generated in a machine, this is not economical. A better design is one in which the active materials are fully utilised, and adequate insulation and cooling systems are provided.

The temperature of the various parts of the machine fluctuate with changes in load, and this may subject the insulation system to additional stresses. The designer must ensure that the insulating materials used in a machine are able to withstand the expected rise in temperature under all operating conditions. The importance of insulation and ventilation system design cannot be overemphasised; as electrical machine failures are predominately due to thermal stresses.

Recent developments of synthetic resins and various plastic materials have made a considerable impact on the sizes of electric machines. In particular, slot liners have been the subject of considerable development in both materials used and their method of insertion. Size reduction is more pronounced in small machines where, relatively, the insulations occupy more space.

1.3.1 Insulation classes

According to the International Standards (IEC 85, 1984), insulating materials are divided into seven classes (Y, A, E, B, F, H and C), each specified with a limiting temperature. A brief summary of the classes of *IEC Publication 85* and of the materials listed as examples is given below.

Class Y, up to 90 °C. Unimpregnated paper, cotton or silk, urea formaldehyde, vulcanised natural rubber, and various thermoplastics limited by their softening points.

Class A, up to 105 °C. Paper, cotton or silk impregnated with oil or varnish, or laminated with natural drying oils and resins or phenol formaldehyde; esterified cellulose fibres, polyamides and a variety of organic varnishes used for wire coating.

Class E, up to 120 °C. Phenol formaldehyde and melamine formaldehyde mouldings and laminates with cellulosic materials; polyvinyl formal, polyurethane, epoxy resins and varnishes; cellulose triacetate, polyethylene terephthalate, oil modified alkyd.

Class B, up to 130 °C. Inorganic fibrous and flexible materials (such as mica, glass or asbestos fibres and fabrics), bonded and impregnated with suitable organic resins;

shellac bitumen, alkyd, epoxy, phenol- or melamine-formaldehyde.

Class F, up to 155 °C. As class B but with resins approved for this class: alkyd, epoxy, silicone-alkyd, and certain others.

Class H, up to 180 °C. As class B but with silicone-resins; silicone rubber. Several other materials are now available, and these are discussed below.

Class C, above 180 °C. Mica, asbestos, ceramics, glass (alone or with inorganic binders or silicone resins), polyimides and polytetrafluoroethylene.

The international seven-class system takes into consideration only one quality of the material; its limiting temperature. Insulating materials, however, will require, in most cases, other properties; for example, tearing and bending strength, smooth surface and ability to lie flat (and at the same time being capable of accepting impregnating agents) and abrasion resistance. This strengthens the argument in favour of a more elaborate classification system. On the other hand, such elaboration could be more confusing than effective.

1.3.2 Conductor insulation

The main requirements of the conductor insulation material are thinness, ease of application, high abrasion resistance and, of course, the ability to withstand thermal stresses. Nowadays, these requirements are met by oil base or various synthetic base varnishes (enamels). Round and rectangular wires may also be insulated by overlapped layers of insulating tapes; such as glass-fibre tapes and aramide paper tapes. However, most of the wire insulation requirements of conductors used in small electric machines are met by enamels. The types and properties of enamels most widely used for covering conductors are given below.

Oleoresinous. Made from refined and treated natural oils (such as linseed or soya bean oil), oleoresinous varnishes (class A) have largely been replaced in recent years by synthetic polymers offering better mechanical properties.

Polyvinyl acetal & polyvinyl formal. Produced by chemical processing, and modified with phenolic resins, these enamels offer the advantages of good abrasion resistance and smoothness; which makes the coated wires particularly suitable for use with automatic winding and inserting machines. These coatings were initially generally in the A class. Nowadays, several wire manufacturers offer class E wires coated with a single coat of improved polyvinyl acetal [1.12].

Polyurethane. Depending on the chemical composition and processing, the upper continuous temperature limit of commercially available polyurethanes range from 105 °C to 155 °C. Polyurethane coatings have good resistance to moisture, various chemicals, petrol, oils and ozone. More important, their good abrasion resistance and ability to solder through are responsible for their widespread use in wire coatings.

Polyester (terephthalate). This material retains its properties up to a temperature of 175 °C, and it is generally offered as a class F material. It has a high mechanical strength and a good abrasion resistance, but its chemical resistance is very low.

Polyimide. This material is suitable for continuous use at between 220–240 °C, and it will survive a few hours at twice these temperatures. It is resistant to all organic solvents and to acids. The use of polyimide coatings is limited to special designs where the material's exceptional properties can be exploited. Polyimide films are more widely used, unsupported or laminated with layers of other materials as a surface insulation material.

Polyesterimide. As the name implies, this material is a hybrid of the poly**ester** and the poly**imide** types, and it has properties (and prices) intermediate between the two types. It is easier to process and has a lower limiting temperature (180 °C) than the polyimides; and it has better mechanical and chemical properties than the polyesters. This material is widely offered as enamels for class H machines.

The synthetic enamels mentioned above are the most widely used in small machines. A full list of enamels used for covering conductors, and their temperature indices, can be found in reference [1.13]. It should be noted, however, that the limiting temperature values of commercially available synthetic materials are now, in most cases, higher than those given in reference [1.13].

1.3.3 Slot insulation

The insulating materials used in a slot of a small electric machine may be identified as: slot liner (sleeve), cover slider (slot closure), slot wedge and, in the case of a multilayer winding, slot separator (intermediate slider). While providing the required thermal and mechanical properties, the volume of the (inactive) slot insulation system must be kept to an absolute minimum, so that the volume of active conducting material in the slot can be maximised. At present, slot insulation requirements of small electrical machines are met by the following mica-free materials: single layer (paper-like materials), composite laminates, prepreg (pre-impregnated) sheets and solid insulation materials. Examples of modern insulating materials are given below.

1.3.3.1 Insulating papers

Class A: Pressboards may be prepared by pressing and heat treating a number of wet layers of cotton rag fibres or pulp processed like other electrical papers. They are produced with a minimum thickness of about 0.1 mm. Impregnated with resins, electrical pressboards are used as slot sleeves (0.25–0.5 mm thick) in some small motors. However, their most important use is as a component in fabricating composite insulation materials. Although these materials are specified with a limiting temperature of 105 °C, they are able to withstand short term exposure to much higher temperatures.

Class B: Sheets of fleece–like material based on polyethylene terephthalate have absorption affinity with impregnating agents; a property which is fully utilised in composite materials used in slot insulation. In woven form and when impregnated with varnish, tapes of this material may be used in the general finishing of the winding.

Class F: The condensation product of ethylene glycol and terephthalic acid is a plastic film known as a polyester film. It has excellent mechanical and thermal properties. Polyester films retain their properties up to a temperature of 175 °C; and they have a temperature index of 155 °C.

Class H: Plastic papers on a base of **ar**omatic poly**amide** (aramide paper), first introduced by the Du Pont Company under the trade name NOMEX®, have exceptional thermal and chemical properties. Aramide papers are widely used in class H applications. Aramide papers are also widely used, in composite form, for class F applications.

Class C: Polyimide films, first introduced by the Du Pont Company with the trade name of KAPTON®, exhibit excellent properties up to a temperature of 220–240 °C. Thin polyimide films with thicknesses of 25 or 50 μm are now widely available as a class C material. They are also used in fabricating a wide range of class H (laminated with aramide paper) and class C (in combination with glass fabric layers) composite materials.

1.3.3.2 Composite materials

Composite laminates are produced by bonding plastic films (polyester or polyimide films) with electrical pressboards, fleece material, fabric or plastic paper (aramide paper). The resulting composite material combines the desired properties of its components; yielding technical and economical advantages. For example, the plastic film contributes excellent mechanical and thermal properties, while the fibre material guarantees good impregnating capacity.

Two- and three-layer composites are now widely available. Normally, two-layer materials are cheaper, but triple-layer materials offer the advantage of reduced hygroscopic effects of the components (pressboard or aramide paper). Plastic films form the core of modern insulation technology, and at least one film is used in multilayer insulators. Polyester films are used to fabricate class B and class F composites, while polyimide films are employed in class H and class C laminates. Depending on the material type, the thickness of the polyester film varies between 25 to 125 μm, while the thickness of the polyimide film is typically 25 or 50 μm. Examples of the composition of triple-layer insulation materials are shown in Figure 1.4 [1.14].

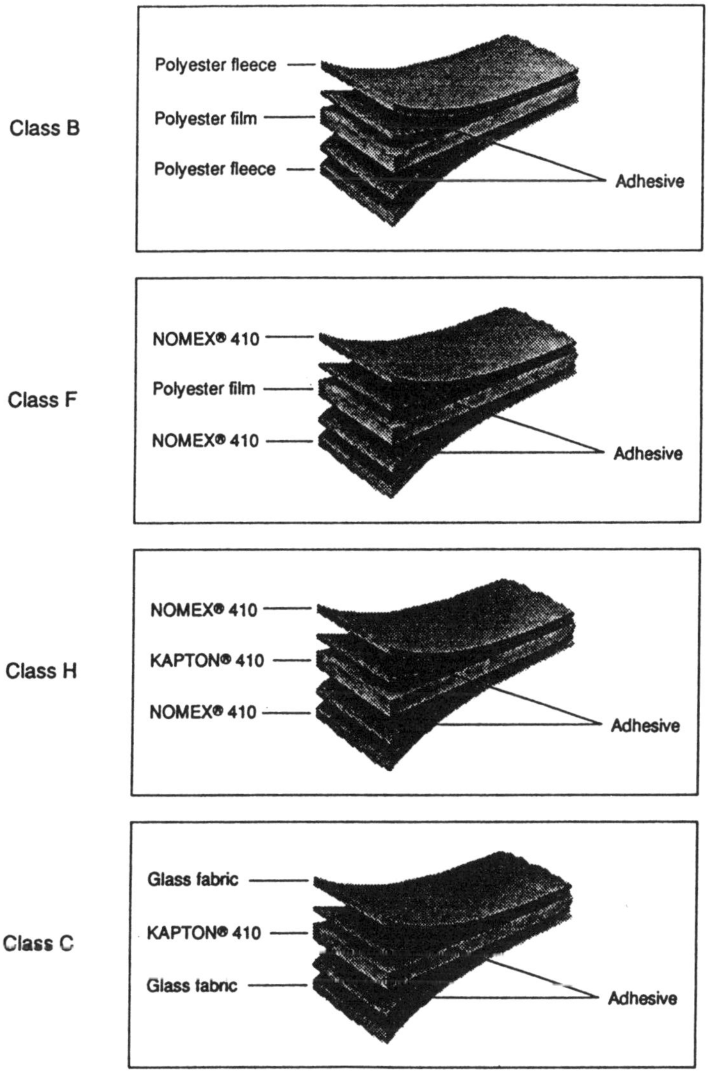

Figure 1.4 Examples of multilayer insulation materials [1.14].

Multilayer insulation materials are widely used in small electric machines as slot sleeves, as cover slider, as slot separator of multilayer windings and as a circumferential insulator in small rotors.

1.3.3.3 Pre-impregnated sheet materials

Preim**preg**nated sheet materials are commonly described by the word prepregs. Prepregs are produced by impregnating or coating a base material (fabric, fleece, aramide paper, plastic film or multilayer insulation material) with insulating resins (such as phenolic resin, epoxy resin, polyesterimide resin or polyimide resin), and partially curing to a dry state. The required thermal and mechanical properties, as well as economic considerations, determine the base material and the resin system to be used.

Prepregs can then be easily transported and stored. However, the main advantage of using prepregs is that the impregnating agent is introduced into the slots or coil winding together with the surface insulation. Pressing conductors into the resinous prepregs causes the resin to flow, filling any existing cavities and resulting in an intimate solid bond between the insulating material and conductors. Heat treatment is normally required to fully cure the resins. Compared with conventional impregnation processes, prepregs can save processing time and energy. This is because the hardening time for prepregs is considerably shorter. Prepregs must be stored in a cool dry place. Depending on its resin-hardener system, a prepreg may be stored for up to 6 months at 20 °C and up to one year at 5 °C.

1.3.3.4 Solid insulation materials

In the past, solid insulation materials have been made from wood, porcelain, glass or mica. Nowadays, solid insulators can be made by consolidating several layers of prepregs under heat and pressure. Plastic sheets of thicknesses between 1 to 100 mm may be produced by this technique. Slot wedges and spacers can be cut from the solid sheet as required. Alternatively, fibre-reinforced plastic profiles may be manufactured directly and more economically as described in reference [1.14].

1.4 PERMANENT MAGNET MATERIALS

Permanent magnets may be used in electric machines to provide excitation; producing the same effects as electromagnets excited with direct current. A permanent magnet is useful because it stores magnetic energy, and this energy is not consumed in the device operation. It plays a role comparable to that of a catalyst in a chemical reaction; it is a necessary intermediary, but its energy is not used up. When operated within normal limits, the magnet retains its energy for an indefinite period of time. It may be noted that if a magnetic field is established using an electromagnet instead of a permanent magnet, the excitation field energy is still conserved. However, some energy is dissipated, namely, the ohmic losses of the excitation current.

1.4.1 Permanent magnet fundamentals

Permanent magnets are hard magnetic materials characterised by large hysteresis loops, as shown in Figure 1.5. When a magnet material is subjected to a magnetising field (i.e. placed between the magnetising poles of an electromagnet), the flux density in the material will increase as illustrated by the curve 0–1 of Figure 1.5; which is known as the initial magnetisation curve. The material saturates at point 1, and

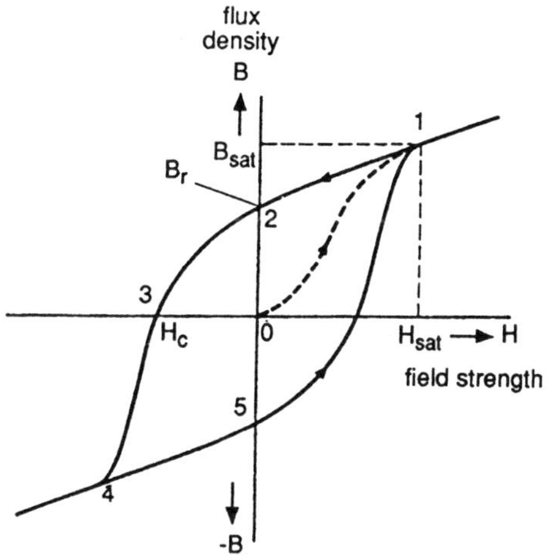

Figure 1.5 Permanent magnet hysteresis loop.

a further increase in the magnetising field (H) will result in a marginal increase in the flux density (B); the flux density increases at a rate approaching the permeability of the free space (μ_0). Steady reduction of the magnetising field, after reaching saturation, causes the B–H locus to follow the curve 1–2. The value of the flux density at point 2 on the hysteresis loop ($H = 0$) is known as the *remanence or residual flux density* (B_r) of the magnet material; and it is indicative of the amount of magnetic flux the material can produce.

Reversing the direction of and increasing the magnetising field produces the second quadrant of the hysteresis loop, curve 2–3; known as the *normal demagnetisation curve* and this is the most important region of the magnet characteristic. The value of the magnetising field at which the flux density in the magnet reaches zero is known as the *coercivity or the coercive force* (H_c).

Further increase in the magnetising field drives the magnet material to saturation in the reversed direction (point 4). The hysteresis loop is completed by reducing the magnetising field to zero until point 5 is reached, and then reversing the applied field again to its original polarities and increasing it until point 1 is reached.

The flux density values used to plot the hysteresis loop of Figure 1.5 represent the total flux density in the magnet material. Not all the flux in the magnet material is due to its inherent (intrinsic) properties. Without the magnet material between the magnetising poles there will still be flux lines in the air-gap. The flux density in an air-gap subjected to magnetising field H is, of course, $\mu_0 H$. Consequently, the total (or normal) flux density in the magnet material (B) consists of two components, one equal to $\mu_0 H$ (which would have existed in the air anyway) and the other is the intrinsic flux density B_i (owing to the material's inherent ability to have more flux with a given applied H than will appear in the air-gap). Arithmetically, $B_i = B - \mu_0 H$ in the first and fourth quadrant, and $B_i = B + \mu_0 H$ in the second and third quadrant of the hysteresis loop; as H has a negative sign in the second and third quadrant. The plot of B_i against H is known as the intrinsic hysteresis

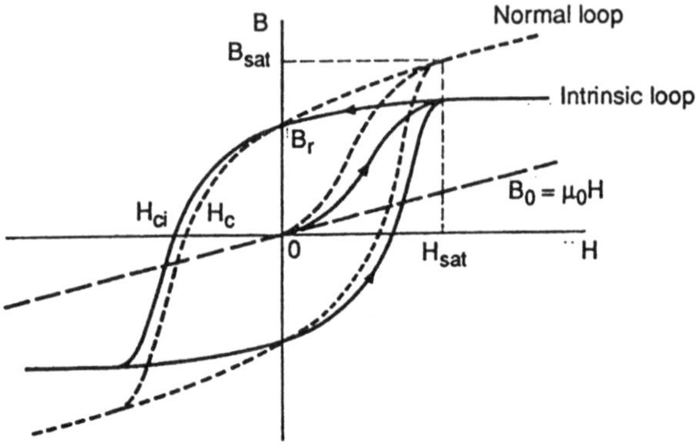

Figure 1.6 Normal and intrinsic hysteresis loops of a permanent magnet material.

loop of the magnet material. Figure 1.6 illustrates the normal and the intrinsic hysteresis loops of a magnet material. We now have two demagnetisation curves, the *normal* and the *intrinsic.*

The residual flux density, or remanence, is the same for both the intrinsic and the normal demagnetisation curves. The coercivity, however, is different.

The *intrinsic coercivity*, H_{ci}, is higher than *normal coercivity*, H_c. The difference between H_{ci} and H_c depends on the slope of the normal demagnetisation curve in the vicinity of H_c; the higher the slope the lower the difference will be. The slope of the demagnetisation curve as it intersects the $-H$ axis is very high for Alnico magnets, and, therefore, there is little difference between the intrinsic and normal coercivities. Ceramics (or ferrites) and rare earth magnets have nearly linear normal demagnetisation characteristics between B_r and H_c, and the difference between H_{ci} and H_c is more significant. In some rare earth magnets H_{ci} is about twice H_c.

1.4.2 Modern magnet materials

Permanent magnet materials can be grouped into three main groups according to their chemical composition. These are: ceramics (or ferrites), Alnicos and rare earth magnets. Ferrites (magnetically hard ceramics) are electrical and thermal insulators while all other magnets are metallic conductors. Alnicos have relatively high remanence and low coercive force, ceramics have low remanence and fairly high coercivity, while both these parameters have high values for rare earth magnets. Ceramics use very cheap and plentiful raw materials. Alnicos and rare earth–cobalt (samarium–cobalt) magnets use cobalt, but with greatly different efficiency, while ceramics and rare earth–ferrites (neodymium–iron–boron magnets) require none at all.

The specifications of permanent magnet materials are subject to the International Standard IEC 404-8-1 (1986). According to *IEC Publication 404-1*, permanent magnet materials are classified by a letter followed by numbers. Alloy magnets are classified by the letter R, while ceramics are identified by the letter S. The first number designates the type of the material in the respective class. For example, R1 designates Alnico magnets and R5 identifies the rare earth–cobalt group. The

Table 1.7 Magnetic properties of magnet materials

	$(BH)_{max}$ (kJ/m^3)	B_r (T)	H_c (kA/m)	Price per unit energy (£/J)
Nd–Fe–B	200–290	1.20	870	2.3
$SmCo_5$	130–190	0.97	750	5.7
Sm_2Co_{17}	180–240	1.05	660	5.1
Alnico	70–85	1.1	130	2.7
Ceramics	27–35	0.4	240	0.3

second number distinguishes between: isotropic magnets (0), anisotropic (1), isotropic polymer bonded (3) and anisotropic polymer bonded (4). The number in the fourth position corresponds to various grades within the same magnet group.

1.4.3 Magnetic properties

The most appropriate single parameter for characterising the quality of a magnet is the maximum energy product $(BH)_{max}$ of its induction and magnetic field; as it represents the maximum energy available from the magnet. The size of the permanent magnet is at a minimum when the magnet is operating at its maximum energy product point, $(BH)_{max}$.

The best high performance permanent magnets were the rare earth–cobalt (SmCo) materials having maximum energy products between 130 and 190 kJ/m^3. The emergence of the cobalt-free neodymium–iron–boron material in 1984 [1.15] with a maximum energy product of 290 kJ/m^3 changed this situation. The pace of developing this new material has been very rapid during the last few years, and now this material is available in commercial quantities from a large number of magnet manufacturers.

Table 1.7 summarises the magnetic properties of the main groups of magnet materials and gives the approximate cost per unit energy of each. There are several grades of each material group; hence the wide range of the maximum energy product. The remanence and coercivity are taken as the average values for the grades commonly used in electric machines. Properties of low energy grades of the Alnico family and polymer-bonded ceramics and neodymium–iron–boron materials are not included in the comparison.

It should be noted that the price of a permanent magnet depends largely on the dimensions, the complexity of the machining to be carried out and tolerances for dimensions and magnetic properties. There are also some variations in price within the same group. The cost per unit energy given in Table 1.7 is an average cost and should be used primarily for comparison purposes.

1.4.4 Thermal properties

When the temperature is increased, some degradation of magnetic properties occurs; some of it reversible, some irreversible. The reversible changes in remanence and

Table 1.8 Temperature effects on magnet materials

	Curie temp. (°C)	α (%/K)	β (%/K)	Max. operating temp.
Nd–Fe–B	310	−0.13	−0.60	140
SmCo	720	−0.045	−0.25	250
Alnico	830	−0.02	+0.01	500
Ceramics	450	−0.20	+0.40	300

coercive force are usually quoted in per cent per Kelvin; described by temperature coefficients α and β respectively. The irreversible changes are functions of temperature, the operating point of the magnet, and the strength of the demagnetising field. Table 1.8 gives the Curie temperature, temperature coefficients and recommended highest operating temperature for the main groups of magnet materials. The data shown are obtained by averaging the relevant values for several grades within each magnet group. At the material's Curie temperature any magnetisation will be lost, and a renewed magnetisation at a lower temperature will be needed. Most materials also exhibit a transition temperature at which the crystal structure changes, resulting in permanent loss of the hard magnetic properties. Obviously, the maximum operating temperature is set below the lower of these two values.

It should be noted that some grades of Nd–Fe–B magnets show an unfavourable behaviour at elevated temperatures, exhibiting high temperature coefficients for both remanence and coercive force. This limits the maximum operating temperature to approximately 140 °C.

1.4.5 The impact of Nd–Fe–B magnets on motor design

New improved permanent magnet materials can have a threefold impact on electromagnetic devices, by:

(1) Replacing other materials in some existing products.

(2) Replacing electromagnets in some existing applications.

(3) Opening up completely new applications for which permanent magnets had not been previously viable.

The replacement of other permanent magnet materials is of considerable importance for the commercial success of a new material. Direct replacement is possible where the demagnetisation curves of the old and new material are parallel, so that the same operating line can be used. Because of the similarities between the demagnetisation curves of Nd–Fe–B and SmCo magnets, the latter are being replaced in several applications. However, SmCo is retained in applications where the magnet is subjected to high demagnetising fields at elevated temperatures during operation.

Of all permanent magnet materials, ceramics offer the least expensive magnetic energy. However, even they are being replaced by Nd–Fe–B magnets in many applications where low volume and weight are required. Additionally, the high energy product of neodymium–iron–boron materials allows the use of smaller magnets. This results in a reduction in the size of other components in the machine, such as iron parts and winding, which may yield a lower overall cost.

Neodymium–iron–boron materials, at present prices, offer less costly magnetic energy than Alnicos and the use of the latter in motors is diminishing. However, by virtue of their low induction temperature coefficients (α), Alnico magnets are used, in measuring instruments.

1.5 REFERENCES

[1.1] Steinmetz, C. P. (1971) *Lectures on Electrical Engineering*, (P. L. Alger, ed.), Vol. 1, Dover Publications, New York.

[1.2] Stephenson, E. T. (1985) *J. Appl. Phys.*, **57**, p. 4226.

[1.3] Moses, A. J. (1990) *Proc. IEE, Pt. A*, **5**, p. 233.

[1.4] Moses, A. J. (1992) *J. Magnetism and Mag. Mat.*, **112**, p. 150.

[1.5] Chen, C. W. (1977) *Magnetism and Metallurgy of Soft Mag. Mat.*, North-Holland, Amsterdam, p. 276.

[1.6] Armco Steel Corporation (1970) *The Metallurgy of Iron and Silicon–iron for Soft Magnetic Applications*, Technical Report.

[1.7] Telcon Ltd, *Cobalt Iron Alloys*, Product brochure, Telcon Ltd, Crawley, Sussex, England.

[1.8] Hamdi, E. S. (1985) *ERA Report No. 85-0087*, ERA Technology Ltd, Leatherhead, Surrey, England.

[1.9] Johnson, L. A. (1983) *Utilisation of Amorphous Metals in Electric Motors*, General Electric Company Report.

[1.10] Jordan, H. E. and Woods, E. J. (1984) *Fractional Horsepower Motor Constructed with Amorphous Iron*, Reliance Electric Company, Cleveland, Ohio.

[1.11] Kienle & Spiess (1985) *Stator and Rotor Laminations for Induction Motors*, Stanz- und Druckgiêwerk GmbH, D-7123 Sachsenheim, Germany.

[1.12] Von Roll Isola (1991) *Winding Wires*, Specification No. 2-1400-50, UDD-FIM S.A., France.

[1.13] Chalmers, B. J. (ed.) (1988) *Electric Motor Handbook*, London: Butterworths, p. 336.

[1.14] August Krempel Soehne GmbH, *Electrical Insulation Materials*, D-7143 Vaihingen/Enz 2, Germany.

[1.15] Sagawa, M. *et al.* (1984) *J. Appl. Phys.*, **55**, No. 6, Pt. 2A, p. 2083.

2

HEATING AND COOLING OF ELECTRICAL MACHINES

2.1 INTRODUCTION

During the process of energy conversion in an electric machine some energy is lost in the form of heat. These losses may be divided into constant losses, which occur at no load, and load-dependent losses which vary approximately as the square of the armature current. The constant losses include iron losses, friction, windage and copper losses associated with the magnetising current. The load-dependent losses are primarily due to the copper losses in the winding.

Cooling is provided to stabilise the operating temperature of various parts of the machine to within limits set by the thermal classification of the insulation materials used.

Each insulation material has a certain operating temperature at which it functions reliably for a sufficiently long period compatible with the expected life of the machine. A relatively small increase of temperature reduces the service life of the insulation material significantly. For example, the service life of some organic class A materials can be expressed by:

$$t = K\,e^{-\alpha\theta} \qquad (2.1)$$

where: t is the service life in years, θ is the operating temperature in °C and K and α are constants. Typically, K is 7.15×10^4 years and α is 0.08 for class A insulation materials. Therefore, such a material can function reliably for 16 to 24 years at temperatures between 100 °C and 105 °C, but its life time would be about seven years if the operating temperature increases to 115 °C. This highlights the importance of accurately predicting the operating temperature of various parts of the machine.

Matters are more complicated with inorganic insulations; such as polyester films. This is mainly because these materials are not usually used in the pure state; they are employed as products containing binders, varnishes, resins, etc. These substances are generally affected to a greater degree than the basic insulating material.

As well as reducing the service life of the insulation system, an excessive temperature rise may also adversely affect the mechanical components. For example, a commutator may lose its regular geometrical shape, soldered joints between the armature coil sides and the commutator may break, bearings may fail, etc. This, again, emphasises the importance of the thermal design.

Thermal design of electrical machines involves not only the design of the insulation system but also the cooling arrangement. In recent years, thermal design of electrical machines has acquired a great deal of interest, and this has enabled better utilisation of active materials without deterioration of performance. This chapter is devoted to general aspects of thermal design of small electrical machines.

2.2 TEMPERATURE RISE, TYPES OF DUTIES AND RATINGS

2.2.1 Temperature rise

The temperature of a given part of an electric machine (θ) depends not only on the losses occurring in it, but also on the ambient temperature, or the temperature of the cooling media (θ_a). The difference between these two values is termed the *temperature rise.*

Various standards give detailed descriptions of temperature measuring methods to be used. In the case of small electric machines, the winding temperature rise is usually determined from the increase in resistance (resistance method). For machines rated at 600 W or less, the standards also permit the use of thermometers or non-embedded thermocouples for measuring the windings and end winding temperature (thermometer method) [2.1].

The temperature rise limits for various classes of machines are established by the international (*IEC Publication No. 34*), and various national standards. These limits are not only specified for the windings, but also for other machine parts which are not generally insulated. Most standards, including IEC and NEMA, use 40 °C as the base ambient temperature.

If the site ambient temperature is higher than 40 °C, the temperature rise limit will be reduced accordingly. On the other hand, the IEC standard permits correction on a degree-for-degree basis up to a maximum of 10 °C for machines operating at ambient temperatures lower than the assumed 40 °C base value. It should be noted that if during certain periods of operation the ambient temperature falls below 40 °C (for example, in winter), this does not mean that the machine may be loaded above its power rating; except in emergencies. This is because the established temperature rise limits are based on the necessity of ensuring normal service life, taking into account the natural variations of the ambient temperature.

The temperature rise limits of motors employed in submerged drives are measured relative to the temperature of the cooling fluid. Provided that the temperature of the cooling fluid is 30 °C or less, the temperature rise limits will be increased by 10 °C (above the values given for air-cooled machines). For higher coolant temperatures correction is made on a degree-for-degree basis.

2.2.2 Types of duties and ratings

There are four classes of ratings:

(1) Maximum Continuous Rating (MCR), defined as the load at which the machine may be operated for an unlimited period while complying with the normal performance requirements.

(2) Short-Time Rating (STR), defined as the load at which the machine may operate for a specified period (usually 10, 30, 60 or 90 minutes) starting at the ambient temperature, while complying with the normal performance requirements.

(3) Equivalent Continuous Rating (ECR), namely the load and conditions at which the machine may be operated without the temperature rise of any part exceeding specified values and while complying with normal performance requirements. The load is considered to be equivalent to one of the standard duty types described below.

(4) Duty-Type Rating (DTR), is the loading condition at which the machine may be operated while complying with the normal performance requirements, in accordance with one of the standard duty types S3 to S8 described below.

The standardised duty types are: continuous running duty (type S1), short-time duty (type S2), intermittent periodic duty (type S3), intermittent periodic duty with starting (type S4), intermittent periodic duty with electric braking (type S5), continuous-operation periodic duty (type S6), continuous-operation periodic duty with electric braking (type S7), continuous-operation periodic duty with related torque/speed changes (type S8) and duty with non-periodic load and speed variation (type S9).

Continuous duty (type S1): On this duty a constant load is applied to the machine for a sufficiently long duration so that thermal equilibrium is reached; thermal equilibrium being defined as the state when the temperatures of the various parts of the machine do not vary by more than 2 °C over a period of one hour. The maximum continuous rating of a machine is defined as the load value at which the machine may operate for unlimited period without the temperature rise of any part exceeding specified values.

Short-time duty (type S2): This involves operation at a constant load for a specified time (N) followed by a rest period (R); the loading period being less than that required for thermal equilibrium to be reached and the rest period being long enough for the machine to return to cold conditions. The short-time rating of a machine is defined by the load and the loading period at which it may operate, starting at ambient temperature, without the temperature rise of any part exceeding specified values. This duty type is indicated by the abbreviation and the duration, for example, S2 60 min.

Intermittent periodic duty (type S3): In this case the loading and rest periods, as defined in connection with short-time duty, alternate. The intensity of loading in this case, and in all other periodic duty cases (duty types S4–S8), is determined in terms of the *cyclic duration factor* (also known as *load factor* or *duty factor*); defined as the ratio of the loading period (N) to duration of the cycle ($N + R$, where R is the duration of the rest period). The cyclic duration factor is generally determined on the basis of a 10 minutes cycle and the recommended values are 15%, 25%, 40% and 60%, for example, S3 25%.

Intermittent periodic duty with starting (type S4): This type of duty consists of identical cycles each containing a run-up period (D), a constant loading period (N) and a rest period (R). At the end of each loading period the machine is stopped by either unforced (natural) deceleration or by means of a mechanical braking method which does not cause additional heating of its windings. The duration of the run-up period is added to the loading period in the calculation of the duty factor. The duty factor is then $(D + N)/(D + N + R)$.

Intermittent periodic duty with electric braking (type S5): Like duty type S4 above, the cycle consists of a starting period (D), constant loading period (N) and a rest period (R). Unlike duty type S4, the machine is made to stop by means of rapid electric braking. Both the starting (D) and braking (F) periods are included in the calculation of the duty factor in this case. Therefore, the duty factor is $(D + N + F)/(D + N + F + R)$.

Continuous-operation periodic duty (type S6): In this case each cycle consists of a period of constant loading (N) and a period of operation at no load (V). The no load period is not included in the calculation of the duty factor which is given by $N/(N + V)$.

Continuous-operation periodic duty with electric braking (type S7): This type of duty consists of identical cycles each consisting of a start-up period (D), a constant loading period (N) and a period of rapid electric braking (F). As there is no rest period (unlike duty type S5), the duty factor in this case is 100%.

Continuous-operation periodic duty with related torque/speed changes (type S8): In this case each cycle consists of two or more periods of operation at different constant loads (N1, N2, etc.) each at a different predetermined speed of rotation. As well as the loading periods, the cycle will of course consist of periods of acceleration (D1, D2, etc.) and braking (F1, F2, etc.). Various cyclic duty factors may be defined in terms of this case [2.1].

Duty with non-periodic load and speed variation (type S9): This duty caters for cases where the machine is frequently overloaded. The load and speed vary non-periodically within the permissible operating range.

2.3 COOLING SYSTEMS AND TYPES OF ENCLOSURES

The subject of cooling and ventilation is closely linked to the type of enclosure used in electrical machines. While the cooling system is responsible for stabilising the temperature of the various parts of the machine within set limits, the enclosure provides its protection against harmful effects. Therefore, the enclosure type is determined by the operating environment. Generally, the higher the degree of protection provided by the machine's enclosure the more sophisticated its cooling system will be.

It is convenient to classify the cooling methods employed in electric machines into direct and indirect cooling. In the forementioned the ambient air is in direct contact with the heat sources (winding and core) and in the latter the heat dissipation is by either conduction through the totally enclosed frame (and by convection from the frame surface to the ambient air) or by an internal cooling circuit operating in conjunction with a heat exchanger. This general classification of the cooling systems yields two main types of enclosures; the open and the totally enclosed.

Direct cooling of open machines generally results in the most economical and simplest design. However, as such arrangements provide little or no protection to the active parts of the machine, open machines are not the most widely used nowadays.

2.3.1 Designation of enclosures

In accordance with international standards (IEC 34-5), the designation used to describe the enclosure type used in an electric machine consists of the letters IP followed by two numerals describing the degree of protection against the ingress of solids and against the harmful ingress of water; respectively. The first digit varies between 0 for non-protected machines and 5 for dust-protected machines. The second digit ranges from 0 for non-protected machines to 8 for machines suitable for continuous operation immersed in water. In some cases the two digits are followed by one of the letters S or M. The suffixes S and M indicate whether the degree of protection was verified with the machine stationary (S) or running (M). When neither letter is included in the IP number, the degree of protection applies to normal operating conditions.

2.3.2 Enclosure types

An open non-protected machine (enclosure type IP00) is one in which there are no restrictions to ventilation other than that stipulated by the integrity of the mechanical construction. Types of open non-protected machines include: (1) an open pedestal machine (has pedestal bearings supported independently of the machine frame) and (2) an open end-bracket machine (having end-brackets of which the bearings form an integral part).

An open protected machine (enclosure type IP11) is a machine in which the internal rotating parts and connections are mechanically protected from accidental contact without impeding circulation of ambient air. A screen-protected machine is one in which protection is provided by a screen of wire mesh or other suitable cover provided with openings. Depending on the size of the openings, three such enclosures are identified: IP21, IP31 and IP41 for machines protected against solid objects greater than 12 mm, 2.5 mm, 1 mm, respectively.

Drip-proof enclosures are similar to screen-protected ones but with ventilation openings situated and protected in such a way that neither solid objects nor water affect the operation of the machine when tilted up to 15° (enclosure type IP22). A spray-proof machine (IP24) is a machine protected against spraying water from any direction.

Weather protected machines are suitable for outdoor installation with ventilating openings that are so designed to minimise the ingress of harmful substances (enclosure type IP44). Weather protected machines may be cooled by providing a continuous supply of ventilating air; in which case the frame will constructed in such a way that air may be conveyed to and/or from the machine through pipes or ducts attached to its casing.

Totally enclosed machines are so constructed that the enclosed air has no connection with ambient air. Totally enclosed enclosure types are IP54, IP55, IP56, IP57, and IP58 for machines protected against splashing water, against water jets, against heavy seas, against the effect of immersion and against continuous submersion; respectively.

Small totally enclosed machines usually rely on heat dissipation by convection from finned or smooth frames. Cooling of a totally enclosed machine (other than submerged machines) is usually augmented by a fan, driven by the machine itself or separately driven, blowing air over its surfaces and through any cooling ducts or holes (when they are incorporated in the machine). When this is the case, the machine is referred to as totally enclosed fan cooled (TEFC). Totally enclosed machines cooled in this way are generally larger and more expensive than equivalent open-ventilated ones.

2.3.3 Designation of cooling methods

There are many variations of cooling arrangements used in electrical machines. According to the international standards (IEC 34-6), the simplified code used to describe the method of cooling employed in an air-cooled electric machine consists of the letters IC followed by two numerals describing the cooling circuit arrangement and the method of supplying power to circulate the coolant, respectively. The first digit varies between 0 for the case of free circulation and 9 when a separately mounted heat exchanger is used. The second digit ranges from 0 for free convection (no power used to circulate the coolant) to 8 for the case of relative displacement (such as when a motor is driving a fan and is cooled by the main air draught). For example, IC00 refers to a machine cooled by free circulation of coolant (first zero) and the movement of coolant is by temperature difference or free convection.

Examples of the more common cooling methods are:

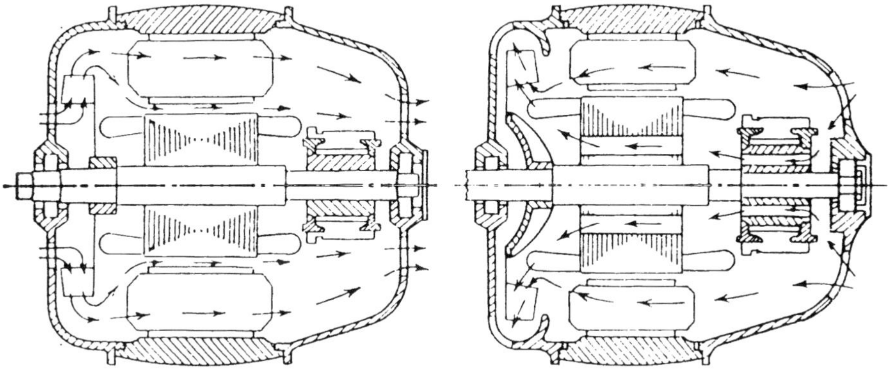

Figure 2.1 Exhaust (left) and forced ventilation (right).

IC01 or IC0	denotes normal ventilation
IC11 or IC1	denotes inlet duct ventilation
IC41 or IC4	denotes enclosed fan-cooled machine
IC05	denotes normal ventilation with separately supplied fan
IC17	denotes inlet duct ventilation with separately supplied fan in the duct
IC48	denotes surface cooling by air stream.

2.3.4 Cooling arrangements

The majority of small electric machines are air-cooled. In conventional machines the temperature of the air coming out rises above that of the entering air by between 18 and 27 °C. According to this, between 0.033 to 0.05 m^3/s (or 2 to 3 m^3/min) of cooling air is required per kilowatt of losses in the machine. The aim of a cooling system is to feed the required quantity of air through a machine with the minimum windage losses and to direct the air to the most heated parts. We now consider the characteristic features of some cooling arrangements relevant to the subject of small machines.

Natural cooling: Naturally cooled machines have no special ventilation system and they are usually of the open enclosure type. This method of cooling is suitable for some small fractional-horsepower motors; since the heating problems are not too severe.

Internal self-cooling: In this case the cooling is augmented by a fan mounted on the machine's shaft and the fan forms an integral part of the machine's rotating components. Depending on the direction of air flow, two types emerge; suction (or exhaust) ventilation and forced ventilation; as illustrated in Figure 2.1. An advantage of suction ventilation is that cold air entering the machine is not heated by the fan losses. These losses could result in an increase of the cooling air temperature by

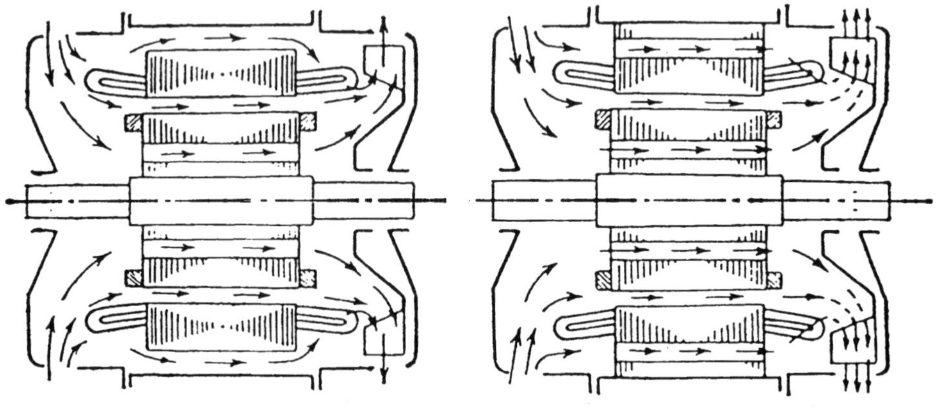

Figure 2.2 Simple (left) and double axial flow ventilation (right).

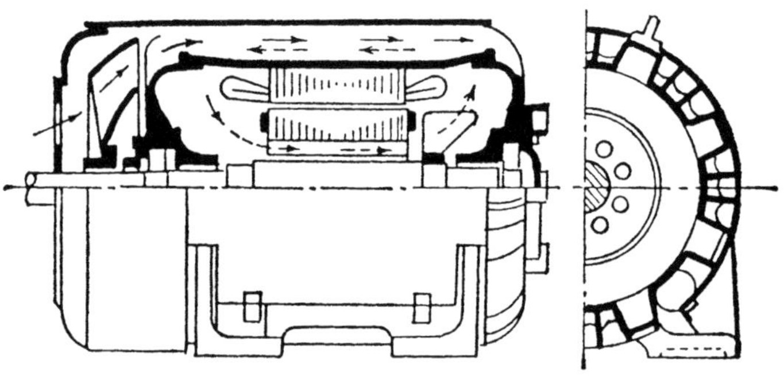

Figure 2.3 A machine with forced external cooling.

3–7 °C; corresponding to an increase of the air delivered to the machine by 15–20%.

Air flow in small electric machines is axial (radial flow is used when cooling ducts are provided in medium and large machines). Axial ducts are sometimes used on the rotor only (simple axial flow) or on both the rotor and stator (double axial flow). Figure 2.2 illustrates the simple and the double axial flow ventilation. A disadvantage of the axial-flow ventilation system is that the heat transfer along the machine is non-uniform. Indeed, the right hand part of the machine in Figure 2.2 is cooled less than the left hand side; a temperature difference in excess of 15 °C is not uncommon.

External self-ventilation: This is used in totally enclosed machines where the active parts, in which losses are produced, are inaccessible to the external ambient air. Therefore, all the heat produced in such a machine can be dissipated only through the external surface of the frame. To increase the rate of heat flow from the inner parts of the machine to the outer surface, the air inside the machine is made to circulate with the aid of an internal fan. The frame, in some arrangements, is cooled by a fan installed behind the end covers. Such an arrangement is termed forced external cooling and is illustrated in Figure 2.3.

2.4 MODES OF HEAT TRANSFER

In general, heat is transferred from one machine part to another, and to the surrounding medium by conduction, radiation and convection. Heat transfer by conduction is significant in electrical machines mainly in solid materials (copper, steel and insulation), whereas heat transfer in air, water and other fluids is mainly by convection. Heat transfer by radiation is generally insignificant as far as electrical machines are concerned [2.2, 2.3]. Therefore, this mode of heat transfer is not considered here.

2.4.1 Heat transfer by conduction

Heat conduction in solids under steady-state conditions is completely described by two laws. Since conservation of energy must hold, the first law states that the divergence of heat flux field is equal to the heat sources in a region. Mathematically, this is expressed by:

$$\nabla\psi = w \tag{2.2}$$

where ψ is the heat flux vector which represents the heat transfer rate per unit area in the normal direction and w is the heat source density. The law of conduction heat transfer states that the heat flux at any point in an isotropic region is proportional to the temperature gradient at the point, therefore:

$$\psi = -\lambda \nabla\theta \tag{2.3}$$

where λ is the thermal conductivity and θ is the temperature. The minus sign is inserted to satisfy the second law of thermodynamics; i.e. heat flows from the higher temperature to the lower temperature points.

For the range of temperatures normally occurring in small electrical machines λ may be assumed constant. Substituting the value of ψ from equation (2.3) into equation (2.2) and remembering that *div grad* is equal to ∇^2, we obtain the following Poisson's equation:

$$\nabla^2\theta = -\frac{w}{\lambda} \tag{2.4}$$

It may be noted that when replacing θ with the electrical potential V, w with the electric charge density ρ and λ with the permittivity ε in equation (2.4) we obtain the differential equation which describes an electrostatic field problem. The similarities between the conduction heat flow and electrostatic field problems can be exploited to facilitate heat flow calculations.

For simplicity, let us consider a one-dimension heat flow problem. Equation (2.3) will be reduced to:

Table 2.1 Thermal conductivity for some materials

Material	λ, W/m °C
Copper	385
Aluminium	205
Electrical steels	20–46
Brass	100
Mica	0.36
Mica tape	0.15–0.6
Varnished cloth	0.2
Cotton cloth	0.07
Pressboard	0.17
Air, 20 °C	0.025
Class A insulation	0.12
Class B insulation	0.0015–0.2

$$\psi = -\lambda \frac{\partial \theta}{\partial x}$$

Assume two parallel surfaces with temperatures θ_1 and θ_2 °C (for instance, a coil side and slot wall), each with an area A (m^2). The amount of heat conducted through the medium between the two surfaces (slot insulation) is given by:

$$Q = \psi A = \frac{\lambda A}{\delta}(\theta_1 - \theta_2) \tag{2.5}$$

where: Q is the heat transferred by conduction, W; λ is the thermal conductivity of the intermediate medium, W/m °C and δ is the distance between the surfaces (insulation thickness), m. Equation (2.5) may be rewritten on the form similar to Ohm's law in an electric circuit as follows:

$$Q = \frac{\theta_1 - \theta_2}{R_{\text{th}}} = \frac{\Delta\theta}{R_{\text{th}}}, \qquad \text{where: } R_{\text{th}} = \frac{\delta}{\lambda A} \tag{2.6}$$

where R_{th} is known as the thermal resistance (°C/W).

The concept of thermal resistance may be applied to cylindrical geometries where the thermal resistance will be:

$$R_{\text{th}} = \frac{\ln(r_o/r_i)}{2\pi\lambda L} \tag{2.7}$$

where r_i and r_o are the inner and outer radii respectively, L is the cylinder's axial length and λ is the thermal conductivity. The thermal conductivity of some materials is given in Table 2.1. If, for example, the losses in a winding are such that 0.5 W is transferred through 100 mm^2 of Class A insulation which is 0.5 mm thick, the temperature drop in the insulation will be:

$$\Delta\theta = \frac{Q\delta}{\lambda A} = \frac{0.5 \times 0.5 \times 10^{-3}}{0.12 \times 100 \times 10^{-6}} = 20.8\,^{\circ}\mathrm{C} \tag{2.8}$$

It should be noted that there are several factors which affect the accuracy of heat flow calculations in electrical machines. For example, the nature of the boundary conditions and the quality of contact at various interfaces have a significant effect on the calculated values of temperature gradients, and this will be discussed later.

2.4.2 Heat transfer by convection

Heat dissipated by convection may be classified into two categories; natural convection and artificial (or forced) convection.

Natural convection occurs because the air near a heated body becomes less dense and, therefore, it rises giving place to cooler air, which in turn gets heated and rises, and so on. Examples of this case are the cooling of the commutator surface in an open-type machine and heat dissipation in naturally cooled machines.

The heat dissipated by natural convection can be calculated by:

$$Q = hA\,(\theta_s - \theta_a) \tag{2.9}$$

where h is the convection heat transfer coefficient, W/m^2 °C; A is the area of the emitting surface, m^2 and θ_s and θ_a are the temperature of the emitting surface and ambient, respectively. A major difficulty arises in determining the convection heat transfer coefficient (h) and, hence, the amount of heat dissipated by convection. This is because the convection heat transfer coefficient is a function of many variables such as the temperature difference between the heated body and the coolant, configuration and orientation of the heated surface and physical properties of the coolant. The values of the convection heat transfer coefficients are, approximately, 6.5 and 890 W/m^2 °C for horizontal cylinders of 50 mm diameter in air and of 20 mm diameter in water, respectively [2.4]. Because the constructional details are different for different machines, it is impractical to give values of convection heat transfer coefficients here. However, the free-convection heat transfer coefficients may be calculated for simple geometries by the formulae given in references [2.5] and [2.6]. As most machines utilise finned structures to facilitate heat removal to the surrounding media, such structures are treated later in this chapter.

In most modern machines heat is removed by artificial, or forced, circulation of coolant. The usual method employed of cooling small machines is by mounting a fan on its shaft. The problem of calculation of heat dissipation by forced convection is very complex; even more so than in the case of free convection. This is because the forced convection heat transfer coefficients depend, in addition to the factors mentioned above, on the velocity of air and on whether the air flow is laminar or turbulent. Therefore, the increase in heat rejection when ventilating open surfaces is taken into account by means of empirical formulae. One of the simplest formulae is:

$$\bar{h} = h\,(1 + C_a v^n) \tag{2.10}$$

where $\bar{h}$ and h are the coefficients of heat transfer from a ventilated surface and for natural convection; respectively, C_a is an empirical constant, v is the air velocity relative to the cooled surface, m/s and n is a constant which has values between 0.5 and unity. More detailed analyses are given later in this chapter.

If the air flow is uniform over the entire surface, then according to experimental data, C_a is about 1.3. When the air flow is not uniform the coefficient C_a drops to a value less than unity.

Equation (2.9) may be conveniently rewritten on the form:

$$Q = \frac{\Delta\theta}{1/\bar{h}A} = \frac{\Delta\theta}{R_c}, \qquad R_c = \frac{1}{\bar{h}A} \tag{2.11}$$

where R_c is the convection heat resistance. This enables the construction of a lumped circuit model for the heat convection process in the machine which can be easily coupled to the heat conduction model (the units of both R_{th} and R_c are °C/W).

2.5 HEAT TRANSFER MODELS

Existing thermal models of electrical machines may be grouped into: (1) models based on numerical solution of the heat flow equations, (2) lumped parameters models and (3) models treating the machine's geometry as concentric cylinders (shell models).

2.5.1 Models based on numerical field solution

Models based on numerical field solution of the heat flow problem feature strongly in the literature. Historically, the finite difference method predates the finite element method. For example, Reichert described the application of the finite difference method to the problem of calculating the temperature distribution in electrical machines in 1969 [2.7]. Armor and Chari utilised a three-dimensional finite element model to study the temperature distribution in a turbine generator [2.8]. Their work considered the detailed geometry of the slots and the temperature dependence of the thermal conductivity (which may not be considered constant when hydrogen cooling is employed). The finite element method was used to solve the heat flow differential equation in cylindrical polar form on a mesh consisting of arc-shaped elements coinciding exactly with the shape of the slots. Transient analysis was also performed by Armor [2.9]. The transient solution was obtained by combining the finite element model and a first order finite difference approximation to the thermal field's time derivative. Figure 2.4 shows the isotherm plots for the 2-pole rotor [2.9]. It may be noted that the isotherms shown in Figure 2.4 are approximately circles in the neighbourhood of the slots; a feature which is exploited in the development of cylindrical or shell models.

Recently, Sarkar *et al.* applied the finite element method to predict the temperature distribution in the stator of an induction motor [2.10]. The mesh used in this work

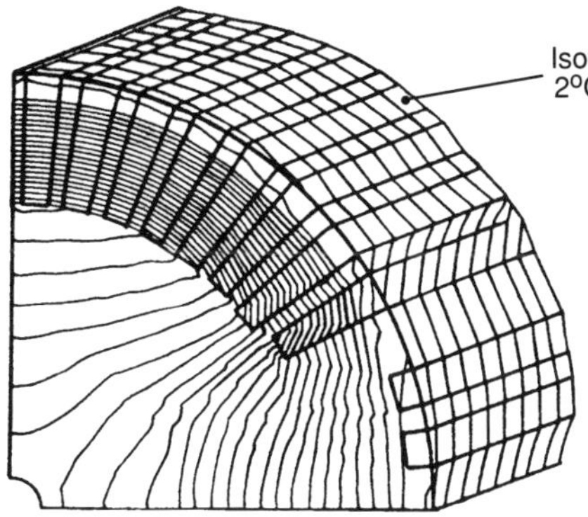

Figure 2.4 Isotherms in a 2-pole turbo-generator [2.9].

utilised arc-shaped elements; as was the case in the work of Armor [2.8, 2.9].

A steady-state thermal model was also proposed by Hamdi *et al.*, and the resultant equation were solved using the finite element method [2.11]. This work, which was concerned with totally enclosed fan-cooled (TEFC) induction motors, demonstrated that the mode of heat transfer in the air-gap is pure conduction for motors running at speeds up to 3600 rev/min. This led to the conclusion that only conduction heat transfer is significant in induction motors supplied at power frequency. Although this model was reduced to a two-dimensional one, a surprisingly good agreement between measured and computed stator winding temperatures was obtained; as is evident in Figure 2.5. The winding temperature was measured by the resistance method [2.11].

2.5.2 Lumped-parameter models

These models exploit the concept of thermal resistance as expressed by equations (2.6) and (2.11) for the conduction and convection heat; respectively. Provided that accurate values for coefficients of thermal conductivity of the various materials and the coefficients of heat transfer from ventilated surfaces can be obtained, this simple concept should yield reasonably accurate results.

A good example to cite regarding the application of the lumped parameters models to heat transfer problems in electric machines is due to Maier [2.2]. This work was mainly concerned with traction motors. As the air flow is entirely internal in traction motors, Maier gives due attention to the problem of determining the air velocity distribution in the various parts of the machine. The work clearly recognises the effect of the air velocity distribution on the values of surface heat transfer coefficients.

Since the heat sink of the motors considered by Maier is the cooling air circulating within the machines, the temperature profile in all the air channel was determined [2.2]. This was obtained as a linear spatial distribution for air flowing along heated surfaces and as exponential distribution for flow in ducts. A system of lumped circuit elements was used to describe the heat transfer process. For example, the

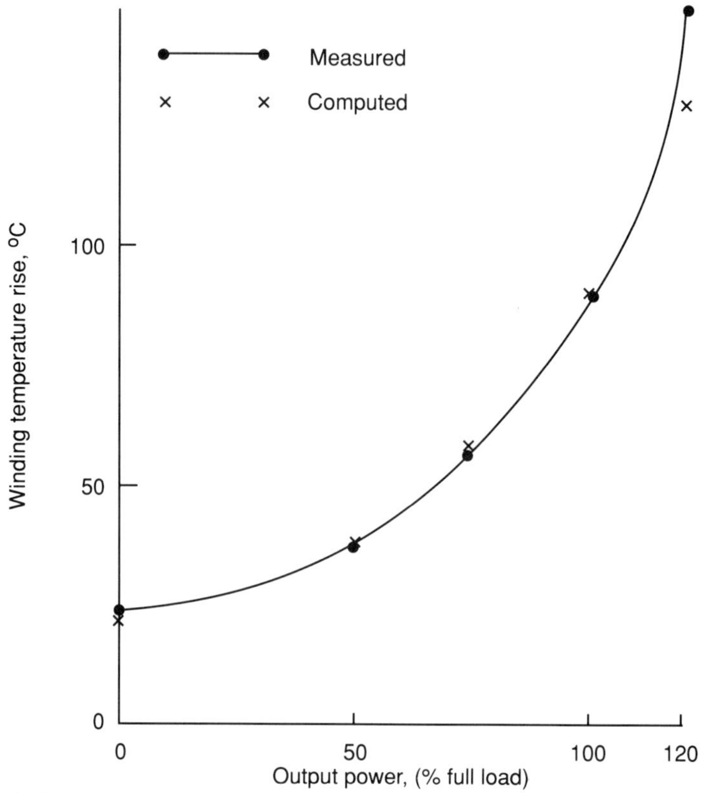

Figure 2.5 Measured and computed stator winding temperature of A TEFC induction motor [2.11].

rotor was divided into eight regions and its equivalent thermal circuit contained eight nodes. The thermal equations were derived in nodal form and each equation contained a recognisable heat source term, a storage term associated with the thermal capacity of the element (in the transient study), a thermal conductance term representing the heat conduction process and, in most cases, a term representing the surface heat transfer to the air coolant. The equations were solved numerically using a first order approximation for the temperature time derivative [2.2].

Among the recent applications of the lumped parameter thermal models to the thermal design problems of industrial induction machines is the work of Mellor *et al.* [2.12]. As is the case in the work of Maier [2.2], the induction motor was divided into regions (10 in Mellor's work) and the resultant thermal network contained an equal number of nodes. Figures 2.6 and 2.7 show the motor's regions and the resultant lumped parameter circuit.

The values of the convection heat transfer coefficients between the frame and external air were determined directly from tests. The free and forced convection heat transfer coefficients were determined from a reduced voltage locked-rotor test and a constant load test, respectively. After thermal equilibrium was reached, the relevant heat transfer coefficient was obtained from the surface-ambient temperature gradient and motor losses as follows:

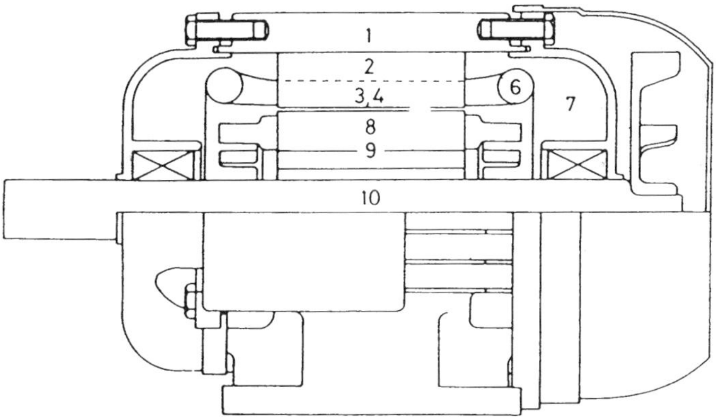

1 Frame
2 Stator
3 Stator teeth
4 Stator winding
5 Air gap
6 Endwinding
7 Endcap air
8 Rotor winding
9 Rotor iron
10 Shaft

Figure 2.6 Thermal regions of an induction machine [2.12].

$$h = \frac{\text{Losses}}{(\theta_s - \theta_a)\,A} \tag{2.12}$$

where θ_s and θ_a are the surface and ambient temperatures; respectively and A is the motor's surface area. Obviously, determination of heat transfer coefficients from tests is only possible for existing machines.

Other values of heat transfer coefficients were calculated using empirical formulae [2.13, 2.14]. An allowance was made for the stator-frame thermal contact resistance.

2.5.3 Shell models

In a shell model the machine is regarded as being made up of concentric cylindrical shells each with different thermal properties, as illustrated in Figure 2.8. This assumption allows the utilisation of much of the work of Taylor [2.13] and Gazley [2.14], which was concerned with the heat flow problem between concentric cylinders, in determining heat transfer data.

Of course, this simplification is only applicable to a certain class of machines where mechanical saliency is absent (such as induction motors and brushless d.c. motors of the surface-mounted magnets type).

A shell model for TEFC induction motors was developed by Sengupta and Sreenivasan [2.3]. Their model consisted of three shells of finite length representing the rotor, stator winding and the core and body lumped together. Like the work of Hamdi *et al.* [2.11], this work concluded that the circumferential air flow in the gap is laminar and therefore the heat transfer process was regarded as pure

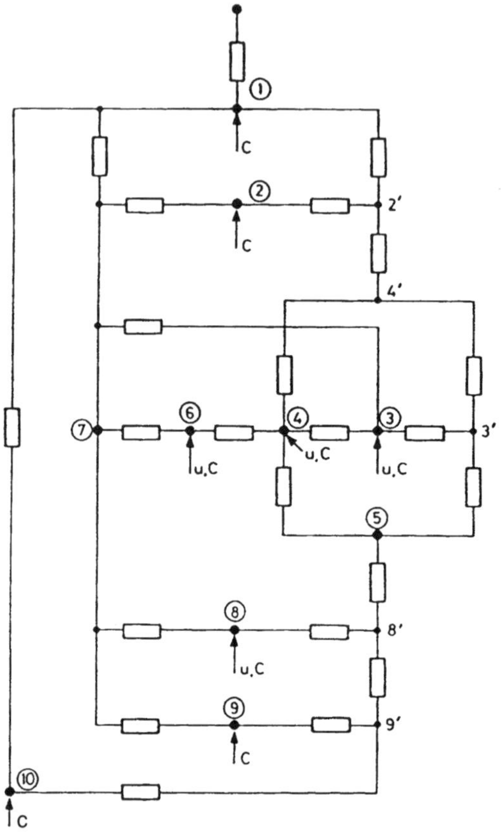

u = nodal heat generation
c = thermal capacitance to ambient

Figure 2.7 Total induction motor thermal network [2.12].

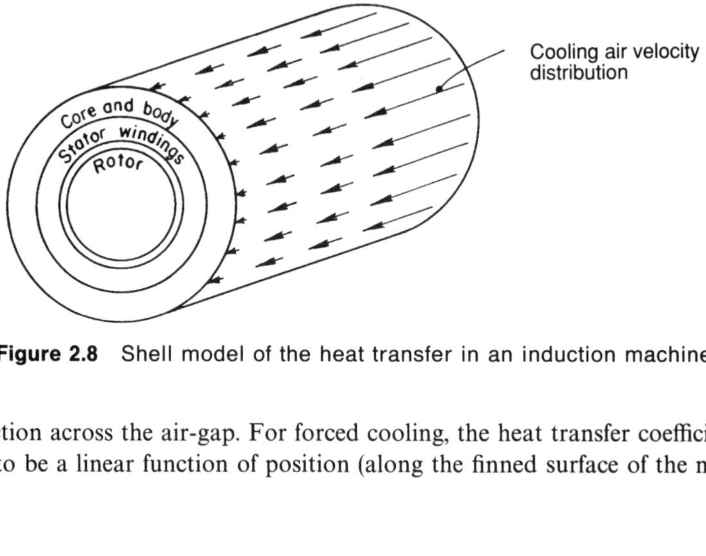

Figure 2.8 Shell model of the heat transfer in an induction machine.

conduction across the air-gap. For forced cooling, the heat transfer coefficient was found to be a linear function of position (along the finned surface of the machine)

Table 2.2 Comparison of calculated and measured temperatures of a totally enclosed fan-cooled induction motor [2.3]. Measured values are shown in parentheses

Region	$z = 0$	$z = 0.25\,L$	$z = 0.5\,L$	$z = 0.75\,L$	$z = L$
Rotor	84.9 (87.0)	86.8	88.4	88.9	88.0 (89.0)
Stator	71.9 (67.5)	70.8 (67.5)	70.3 (67.5)	70.8 (67.5)	71.9 (67.5)
Body	22.5 (19.0)	27.5 (24.0)	34.0 (35.0)	43.0 (41.0)	37.0 (35.0)

[2.3]. This results in the heat flux having axial components as well as radial components. Thus, the steady-state heat transfer differential equations contain only two space variables, and can be readily solved by expanding the temperature as an infinite series in Bessel functions [2.3]. The computed temperatures have been shown to correlate well with those measured [2.3]. The steady-state temperature variation with the axial position are reproduced in Table 2.2 for one of the motors analysed by Sengupta and Sreenivasan [2.3]; where the measured values are shown in brackets.

The data given in Table 2.2 shows that the errors in calculations are largest (15% maximum) in the body region. This is perhaps only to be expected since it is at the surface of the machine where the greatest uncertainties regarding the boundary conditions exist. The calculated (and measured) temperature values in the stator region show a surprising symmetry about the central radial plane of the motor. One would expect a temperature gradient between the hot and cool ends; in particular, when the rotor temperatures show a gradient.

2.6 SOLUTION OF HEAT TRANSFER PROBLEMS

Having discussed the available models which have been applied to solve the heat transfer problem in electrical machines, an attempt is made in this section to help the reader to decide which of these approaches is most suitable for a given device. A general thermal model allows the prediction of the temperature distribution throughout the machine under specified operating conditions. The designer, however, is more interested in thermal indicators such as the temperature rise of the windings. Therefore, a thermal model should be constructed primarily to provide the indicators of interest in a particular application and it should be able to relate these indicators to the main design parameters (such as dimensions) as well as the operating conditions.

A thermal model is regarded as a description of the thermal state and processes in the actual machine, and it may be regarded as three identifiable components, a physical model, a mathematical model and a numerical model.

The starting point is to construct a physical model describing the problem. In such a model one should attempt to simplify the problem as far as it is practically possible without sacrificing any of the important physical features of the device. For example, one such physical assumption could be that the lamination stack may be represented by a homogeneous body with anisotropic thermal conductivity.

Perhaps the most important assumption would be regarding the nature of the air flow in the gap (whether it is laminar or turbulent).

The mathematical model is the formal expression of the physical laws resulting from the physical assumptions made. For example, if the air flow in the gap is laminar, the heat transfer mode across the gap will be pure conduction and the resultant differential equation, which is a part of the mathematical model, should reflect this.

The mathematical model must be distinguished from the various approximation schemes such as lumped circuit, finite difference and finite element which are tools for solving the differential equations. These are termed numerical models.

In many cases numerical models are implemented on computers, but this is not essential to the concept. After all, most of the results of the early work must have been obtained using hand calculators or even slide rules [2.5, 2.13–2.16].

Generally speaking, spending time and effort in developing an elaborate physical and mathematical models for a specific device would result in a relatively simple numerical model being required to obtain solutions. For example, it would take longer to develop an accurate lumped parameter model for a machine than to utilise a finite element package to solve the differential equations describing the heat flow problem in three dimensions. However, the hardware and software requirements of the two numerical models are quite different. A shell model assumes the machine to consist of concentric cylinders; an assumption which in most cases introduce errors. However, an analytical solution of the resulting equations may be obtained; and this eliminates numerical errors associated with numerical solutions. Therefore, designers decide which model to use on the basis of the tools available to them as well as the complexity of the cooling arrangement employed in the machine.

2.7 HEAT TRANSFER DATA

2.7.1 Heat transfer in the air-gap

Because the stator and rotor act as two concentric cylinders rotating relative to each other, earlier work concerned with such geometries may be readily applied to determine the air-gap surface heat transfer coefficient [2.13–2.15]. However, these earlier studies were concerned with smooth cylinders. The heat transfer across the air-gap of an electric machine will be affected by air disturbances caused by the presence of slots. Experimental results suggest that an increase of the heat transfer coefficient by about 10% will be caused by the slots [2.14].

The mode of heat transfer in the gap depends on whether the air flow is laminar or turbulent, and this in turn depends on the radial air-gap length, rotor speed and the ratio of rotor radius to the gap length.

A criterion for transition from laminar to turbulent flow is given in reference [2.15]; only laminar flow exists in the air-gap when the following is satisfied:

$$R_e \leq 41.1\sqrt{\frac{r_1}{l_g}}, \qquad R_e = \frac{l_g v}{\nu} \tag{2.13}$$

where R_e is the gap Reynolds number, r_1 is the rotor radius (m), l_g is the radial length of the air-gap (m), v is the rotor peripheral speed (m/s), and ν is the kinematic viscosity of air (m^2/s).

The air-gap surface coefficient may be determined from:

$$h = \frac{N_{Nu}\lambda}{l_g} \tag{2.14}$$

where λ is the thermal conductivity of air, W/m °C and N_{Nu} is the dimensionless Nusselt number. Nusselt number is given by [2.16]:

$$N_{Nu} = 2.2 \qquad R_e \leq 41.1 \tag{2.15}$$

$$N_{Nu} = 0.23\, R_e^{0.63}\, P_r^{0.23} \qquad 41.1 \leq R_e \leq 100 \tag{2.16}$$

where P_r is the Prandtl number defined as:

$$P_r = \frac{C_p \nu}{\lambda} \tag{2.17}$$

Where C_p is the specific heat of air, kJ/kg °C.

Assuming a line frequency supply, two-pole machines up to frame size 180 would have mainly laminar flow in the air-gap provided its radial length is less than 1 mm. This covers most small industrial induction and synchronous machines and brushless d.c. motors.

2.7.2 Heat transfer in ducts

Radial and/or axial ducts may be incorporated in the design of an electric machine to improve its cooling. However, the use of radial ducts is not a common practice in small machines.

Data on heat transfer in pipes or ducts can be found in a number of textbooks [2.4, 2.17, 2.18]. This is usually presented either graphically or in the form of empirical relationships. Data relating specifically to ducts in electric machines may be found in references [2.19] and [2.20].

2.7.3 Heat dissipation from the surface of the machine

The area of the heat-dissipating surface of an electrical machine is usually increased, without substantially increasing the machine's size, by utilising fins on the frame's outer surface. Usually fins of rectangular profiles are used, but sometimes fins of trapezoidal or parabolic profiles are utilised. Examples of these shapes are shown in Figure 2.9.

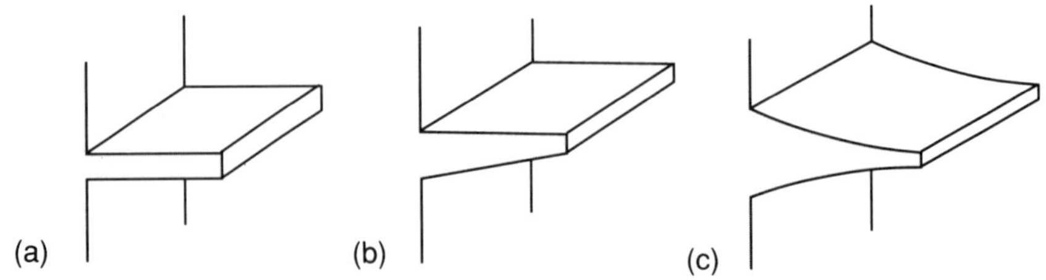

Figure 2.9 Typical fin shapes: (a) a rectangular fin, (b) a trapezoidal fin, (c) a parabolic fin.

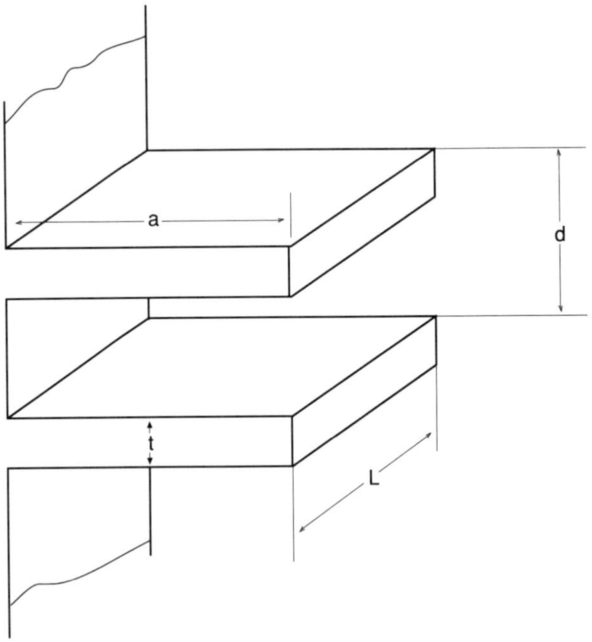

Figure 2.10 Hydraulic channel formed by two adjacent fins.

The modes of heat transfer involving a finned frame and its surroundings may be separated into conduction within each fin and convection (and in the general case, radiation) between the fins and surrounding air.

A thorough mathematical treatment of the "extended surface" heat transfer problem can be found in reference [2.21]. The problem is considerably simplified by assuming that the properties of surrounding air remain constant over the range of temperatures experienced in electrical machines.

A formula for calculating the heat transfer coefficient of the machine's surface assuming constant air temperature is described below. These are based on the method of Heiles [2.22]. In this method the heat transfer coefficient for the finned area is calculated assuming a hydraulic channel being formed by two adjacent fins and a section of the base; as illustrated in Figure 2.10. The equivalent hydraulic diameter of the channel (D) is given by:

$$D = \frac{2ad}{a + d} \tag{2.18}$$

where a and d are defined in Figure 2.10 and their values in equation (2.18) are expressed in metres.

The heat transfer coefficient assuming laminar air flow over the surface is given by [2.22]:

$$h = 284.55\left(\frac{Dv}{L}\right)(1 - e^{x}) \tag{2.19}$$

where

$$x = \frac{0.015L^{0.95}}{D^{1.16}v^{0.21}} \tag{2.20}$$

where L is the length of the fin, m (Figure 2.10) and v is the velocity of the cooling air, m/s.

The expression for the heat transfer coefficient given in equation (2.19) was obtained under laminar flow conditions. When the flow is turbulent the heat transfer coefficient increases by a factor between 1.7 and 1.9 [2.22]. This factor is represented by δ in equation (2.22) below.

In deriving the expression for the heat transfer coefficient, equation (2.19) above, the fin was assumed to have a uniform temperature equal to that at its base [2.22]. In practice, a temperature gradient exists from the root to the tip; the root being at a higher temperature. This results in a reduction of the amount of heat transferred. Therefore, a correction factor (β) was introduced as follows [2.22]:

$$\beta = \frac{\sqrt{t}}{\gamma a} \tanh \frac{\gamma a}{\sqrt{t}} \tag{2.21}$$

where t and a are defined in Figure 2.10, and γ is defined by:

$$\gamma = \sqrt{\frac{2\delta h}{\lambda}} \tag{2.22}$$

where λ is the thermal conductivity of the fin material, W/m °C.

The correction factor β is sometimes referred to as the *fin efficiency* and defined as the ratio of the actual heat dissipation of a fin to its ideal heat dissipation if the entire fin surface was at the same temperature as its base.

The convection heat transfer coefficients of the finned and flat sections of the machine casing are $\beta\delta h$ and δh; respectively.

It may be noted that the analysis presented above is subject to a number of simplifying assumptions. Some of these assumptions, such that a perfect bond exists between the surface and the fin and that no heat is dissipated through the end or through the tip of the fin, are valid for electrical machines. On the other hand, it is assumed that the heat transfer coefficient is constant along the axial length of

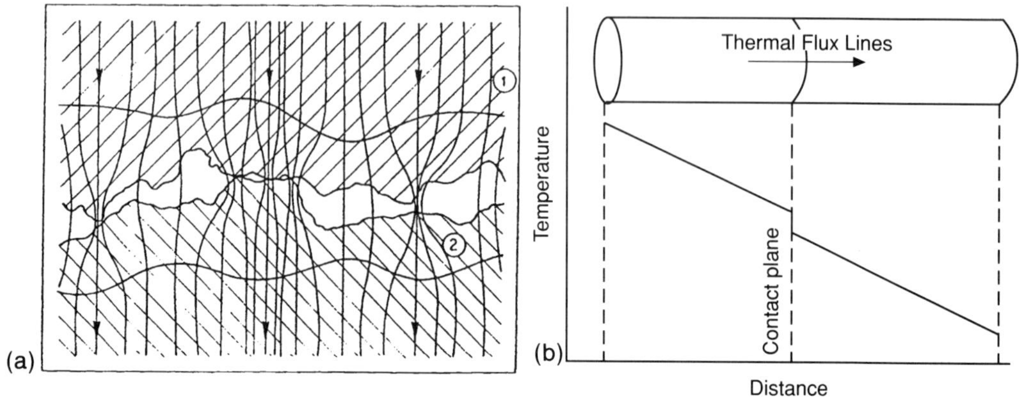

Figure 2.11 (a) Thermal field distribution at an interface (b) and interfacial thermal resistance reflected as a temperature drop.

end. Matters are complicated further by the presence of the terminal box on the casing; which undoubtedly affects the air velocity pattern. However, the analysis presented above should be useful at the initial design stage. After a number of prototype machines are built, these should be tested to provide additional data which may then be used to modify the equations given above for the geometries of interest to a particular designer.

So far we have considered the cylindrical part of the machine's surface. Some heat will be dissipated from the surface of the end covers. The end-cover heat transfer coefficient may be expressed by [2.19]:

$$\bar{h} = h(1 + 0.3v) \tag{2.23}$$

where h is the free-convection heat transfer coefficient and v is the velocity of the cooling air, m/s.

The overall heat dissipation may be obtained by summing the products of the surface area, heat transfer coefficient and temperature difference for each section of the machine casing.

2.7.4 Stator core-frame contact resistance

When two surfaces are brought together to form an interface, the true solid-to-solid contact area between them is generally a small fraction of the apparent contact area. When a uniform temperature gradient is applied along adjoining members, the nett effect of the interface on the heat flow process is the formulation of a temperature discontinuity as shown in Figure 2.11. The thermal flux lines tend to converge to the direct solid-to-solid contact points; since this flow path offers considerably less resistance than the air-filled areas.

The contact resistance to heat transfer is defined as the ratio of the temperature drop at the interface divided by the thermal flux crossing it. Mathematically:

$$R_{co} = \frac{1}{h_{co}} = \frac{\Delta\theta}{\psi} \tag{2.24}$$

where R_{co} is the thermal contact resistance (m^2 °C/W), h_{co} is the thermal contact conductance (W/m^2 °C), $\Delta\theta$ is the temperature drop across the interface (°C), and ψ is the interfacial heat flux density (W/m^2).

The stator core-to-frame thermal contact resistance impedes the heat flow between the main heat sources (winding and core) and the heat sink (ambient air). Therefore, it has an appreciable effect on the temperature rise in electrical machines.

The thermal contact resistance is a function of the physical and mechanical characteristics of the two surfaces, the compressive force with which they are held together and the average operating temperature [2.23–2.25]. It may be estimated from the original dimensions of the frame and stator bore (before the frame is shrunk on to the core) and the physical properties of their materials [2.23]. However, in view of the different manufacturing practices adopted by different electric machine manufacturers, the actual degree of fit between the core and frame is subject to wide variations. Slight variations exist between the precise dimensions of even supposedly identical machines and between various contact locations within a single machine (after all no two machines, even repeats, are exactly alike). This results in a substantial difference between calculated and actual values of thermal contact resistances.

Measured values of the thermal contact resistance between laminated blocks (representing the stator core) and solid mild steel blocks (representing the frame) under various conditions may be found in reference [2.25]. Five blocks of laminations with clamping pressure values of 0.28, 0.42, 0.56, 0.70 and 0.84 N/mm^2 (40, 60, 80, 100 and 120 psi) were used in this work [2.25]. Measurements were obtained for the following surface conditions of the blocks: (1) clean surface as stacked, (2) corroded surface and (3) clean surface machined after stacking. The contact pressure between the laminated and solid blocks (core and frame) was varied between contact weight (regarded as zero) and 0.69 N/mm^2 (100 psi).

The variation of the thermal contact resistance with operating temperature was examined and was found to be insignificant (about 1.5%) over the temperature range of 70 to 130 °C [2.25]. Although a controlled penknife test suggests that a core becomes sufficiently tight when subjected to an axial pressure of 0.28 N/mm^2 (40 psi) [2.26], Hamdi's work [2.25] showed that the thermal contact resistance decreases with an increase with clamping pressure. This effect is more appreciable at low values of contact pressure between core and frame (0.2 N/mm^2 or 30 psi), and it is of primary importance in the case of corroded laminations.

For contact pressures between 0.35 and 0.70 N/mm^2 (50 and 100 psi), it was found that corrosion results in an increase in the thermal contact resistance values of, approximately, between 30% and 80% compared with the values obtained with a clean (unmachined) surface over a wide range of clamping pressure [2.25]. This is an important conclusion bearing in mind that it is not uncommon for machine manufacturers to store fabricated cores for relatively long periods prior to fitting them on to frames.

Machining the core surface was found to reduce the thermal contact resistance by between 18 and 35%, and this effect was found to be, almost, independent of

Table 2.3 Contact resistance values ($R_{co} \times 10^4$) with 0.28 N/mm^2 (40 psi) clamping pressure [2.25]

Contact pressure (N/mm^2)	Clean surface	Corroded	Machined
0.0	24.5	37.3	20.1
0.7 (100 psi)	15.1	28.3	12.4

Table 2.4 Contact resistance values ($R_{co} \times 10^4$) with 0.56 N/mm^2 (80 psi) clamping pressure [2.25]

Contact pressure (N/mm^2)	Clean surface	Corroded	Machined
0.0	20.1	25.8	13.9
0.7 (100 psi)	14.4	21.3	10.6

Table 2.5 Contact resistance values ($R_{co} \times 10^4$) with 0.84 N/mm^2 (120 psi) clamping pressure [2.25]

Contact pressure (N/mm^2)	Clean surface	Corroded	Machined
0.0	19.2	20.1	12.7
0.7 (100 psi)	13.6	18.8	8.5

the clamping pressure [2.25]. This effect is not thought to be significant enough to justify the additional processing step in the production.

Tables 2.3, 2.4 and 2.5 give comparison of thermal contact resistance values for various surface conditions with clamping pressures of 0.28, 0.56 and 0.84 N/mm^2 (40, 80 and 120 psi), respectively [2.25].

For example, let us assume a 0.75 kW, 2-pole machine (frame size 80), the interface area would be about 0.0226 m^2. Assuming an efficiency of 75% and that 90% of the total losses cross the core–frame interface, the interfacial heat flux density would be 7460 W/m^2. If the core clamping pressure is 0.28 N/mm^2 (40 psi) and a contact pressure between the core and frame is 0.7 N/mm^2 (100 psi), we can use the data given in Table 2.3 to estimate the temperature gradient at the interface. $\Delta\theta$ will be about 9 °C, 11 °C and 21 °C for a machined core, an unmachined clean core and a corroded core, respectively. This example demonstrates how values from Tables 2.3–2.5 can be used in the initial design stage to calculate the interfacial temperature gradient until the designer accumulates enough of her/his own data which will, undoubtedly, be more relevant to a particular manufacturing practice and any special characteristic unique to a particular class and size of machines.

2.7.5 Thermal conductivity of solid materials

Data regarding thermal conductivity of winding, solid and laminated core materials are generally available in many textbooks concerned with the general subject of heat transfer. Data regarding most materials of concern to an electric machine designer are presented earlier in this chapter (Table 2.1).

2.8 REFERENCES

[2.1] British Standards Institution (1987) General Requirements for Rotating Electrical Machines — Specification for Rating and Performance, *BS 4999: Part 101.*

[2.2] Maier, F. (1966) *Brown Boveri Review*, **153**, No. 9, p. 574.

[2.3] Sengupta, D. P. and Sreenivasan, V. K. (1977) IEEE PES Winter meeting, New York.

[2.4] Holman, J. P. (1990) *Heat Transfer*, McGraw-Hill Book Company.

[2.5] Bayley, F. J. (1955) *Proc. Inst. Mech. Eng.*, **169**, No. 20, p. 261.

[2.6] Morgan, V. T. (1975) The Overall Convective Heat Transfer from Smooth Circular Cylinders (in *Advances in Heat Transfer*, T. F. Irvine and J. P. Hartnett, eds), Vol. 11, Academic Press, New York.

[2.7] Reichert, K. (1969) The Calculation of the Temperature Distribution in Electrical Machines with the Aid of the Finite Difference Method, **EGZ-A Bd. 90**, H6, p. 137.

[2.8] Armor, A. F. and Chari, M. V. K. (1976) *IEEE Trans.*, **PAS-95**, p. 1648.

[2.9] Armor, A. F. (1980) *IEEE Trans.*, **PAS-99**, p. 934.

[2.10] Sarkar, D. (1991) *Proc. IEE*, Pt. B, **138**, No. 2, p. 75.

[2.11] Hamdi, E. S. *et al.* (1985) *ERA Report No. 85-0123*, ERA Technology Ltd, Leatherhead, Surrey, England.

[2.12] Mellor, P. H. *et al.* (1991) *Proc. IEE*, Pt. B, **138**, No. 5, p. 205.

[2.13] Taylor, G. I. (1935) *Proc. Royal Soc.*, **159**, Pt. A, p. 546.

[2.14] Gazley, C. (1958) *Trans ASME*, p. 79.

[2.15] Taylor, G. I. (1923) *Phil. Trans of Royal Soc.*, **A223**, p. 289.

[2.16] Gazley, C. (1952) *Air Gap Heat Transfer in Rotating Electrical Machines*, GE Report, DF 52 GL 156.

[2.17] McAdams, W. H. (1954) *Heat Transmission*, New York: McGraw-Hill Book Company.

[2.18] Kays, W. M. and London, A. L. (1964) *Compact Heat Exchangers*, New York: McGraw-Hill Book Company.

[2.19] Luke, G. E. (1926) *AIEE Trans.*, p. 1036.

[2.20] Borisenko, A. I. (1965) *Electrotechnika*, **36**, No. 10 (translation in *Soviet Electrical Engineering*).

[2.21] Kern, D. O. and Kraus, A. D. (1972) *Extended Surface Heat Transfer*, New York: McGraw-Hill Book Company.

[2.22] Heiles, F. (1952) *Design and Arrangement of Cooling Fins*, translated in *Engineers Digest*, **13**, No. 10, p. 343.

[2.23] Roberts, D. (1986) The Application of an Induction Motor to Motor Protection and Other Functions, PhD Thesis, University of Liverpool, England.

[2.24] Brunot, I. W. and Buckland, F. F. (1950) *Trans ASME*, p. 253.

[2.25] Hamdi, E. S. (1983) *ERA Report No. 83-0167*, ERA Technology Ltd, Leatherhead, Surrey, England.

[2.26] Walker, J. H. *et al.* (1964) *Proc. IEE*, Vol. **111**, No. 3, p. 565.

2.9 BIBLIOGRAPHY

Bergles, A. E. and Simmonds, R. R. (1971) 'Combined forced and free convection for laminar flow in horizontal tubes with uniform heat flux', *Int. J. Heat Mass Transfer*, ibid, **15**, p. 2535.

Churchill, S. W. and Chu, H. H. S. (1975) 'Correlating equations for laminar and turbulent free convection from a horizontal cylinder', *Int. J. Heat Mass Transfer*, **18**, p. 1049.

Depew, C. A. *et al.* (1975) 'Combined free and forced convection in horizontal, uniformly heated tubes', *ASME Paper 75-HT-19.*

Dippery, D. F. and Sabersky, R. H. (1963) 'Heat and momentum transfer in smooth and rough tubes at various prandtl numbers', *Int. J. Heat Mass Transfer*, **6**, p. 329.

Goldstein, R. J. and Chu, T. Y. (1969) 'Thermal convection in a horizontal layer of air', *Prog. Heat Mass Transfer*, **2**, p. 55.

Habne, E. W. P. (1969) 'Heat transfer and natural convection patterns on a horizontal circular plate', *Int. J. Heat Mass Transfer*, **12**, p. 651.

Husar, R. B. and Sparrow, E. M. (1968) 'Patterns of free convection flow adjacent to horizontal heated surfaces', *Int. J. Heat Mass Transfer*, **11**, p. 651.

Seban, R. A. and Shimazaki, T. T. (1951) 'Heat transfer to a fluid flowing turbulently in a smooth pipe with walls at constant temperature, *Trans ASME*, **73**, p. 803.

Shah, R. K and London, A. L. (1978) *Laminar flow: Forced Convection in Ducts*, Academic Press, New York.

3

GENERAL CONCEPTS AND CONSTRAINTS OF MACHINE DESIGN

3.1 INTRODUCTION

In this chapter we consider the relationship between the rating of rotating electrical machines and their main dimensions, and develop a few general equations which are applicable to all types of rotating machines. However, it must be emphasised that design is a complex process and many factors which affect the design of different types of machines cannot be incorporated in a few general equations. The design concepts and constraints introduced here demonstrate in a general manner how the size of a machine is related to its rating. The detailed design procedures of different types of machines are given in the following chapters.

3.2 SYMBOLS AND DEFINITIONS

The various symbols used in this section are as follows:

D armature diameter or stator bore, m;

L stator core length, m;

n rated speed, rev/s;

n_s synchronous speed, rev/s;

p number of poles;

a number of parallel paths;

τ pole-pitch, m;

Z total number of armature or stator conductors;

T_{ph} number of turns in series per phase;

I_z conductor current, A;

I_a armature current, A;

I_{ph} phase current, A;

E_{ph} rms value of induced emf per phase, V;

E_g generated voltage (or back emf) of d.c. machines, V;

K_w winding factor;

P machine rating, kW;

f supply frequency, Hz;

P_a power developed by the armature, kW;

Q machine rating, kVA;

ϕ magnetic flux per pole, Wb.

The armature diameter (or stator bore), D, and armature (or stator) core length, L, are known as the *main dimensions* of the machine.

The total magnetic flux in the air-gap of the machine is called the *total magnetic loading*. The average flux density over the air-gap of the machine is known as *specific magnetic loading* (B_{av}). Therefore:

$$\text{Total magnetic loading} = p\phi \quad \text{Wb} \tag{3.1}$$

and

$$B_{av} = \frac{p\phi}{\pi DL} \quad \text{T} \tag{3.2}$$

The sum of the ampere conductors around the armature (or stator) periphery is called the *total electric loading*. The number of armature (or stator) ampere conductors per metre of armature (or stator) periphery at the air-gap is known as *specific electric loading* (ac). Therefore:

$$\text{Total electric loading} = I_z Z \text{ Ampere conductor} \tag{3.3}$$

and,

$$ac = \frac{I_z Z}{\pi D} \quad \text{Ampere conductor/m} \tag{3.4}$$

The first step in the design of rotating electrical machines is to obtain the D^2L product, which is proportional to the volume of the machine, by choosing suitable values of specific loadings.

3.3 OUTPUT EQUATIONS

The equation which describes the relationship between the machine output, its main dimensions, its speed and specific magnetic and electric loadings, is known as the output equation.

3.3.1 Output equation of d.c. machines

The power developed by the armature of a d.c. machine, P_a, can be obtained from the armature generated emf, E_g, and its current, I_a, as follows [3.1, p. 329]:

$$P_a = \left(\frac{P}{a}\phi Zn\right)(I_a) \times 10^{-3} \quad \text{kW} \tag{3.5}$$

Noting that the conductor current, I_z, is equal to I_a/a, equation (3.5) may be rewritten as:

$$P_a = (p\phi)(I_z Z)n \times 10^{-3} \tag{3.6}$$

The term $(p\phi)$ in equation (3.6) is the total magnetic loading, and the term $(I_z Z)$ is the total *electric* loading. Therefore, the output of the d.c. machine is proportional to the product of its total magnetic and electric loadings (which is also true for a.c. machines).

In order to be able to apply the output equation in obtaining designs in which the active materials (iron and copper) are effectively utilised, it is necessary to express equation (3.6) in terms of specific loadings. After all, an absolute value of the total magnetic loading in a machine does not provide any information regarding the extent to which the iron parts are utilised. On the other hand, a value of the maximum flux density in the iron parts of the machine can be readily related to the saturation flux density of the material used; thus providing a measure of the extent to which the material is utilised.

Substituting the values of $(p\phi)$ and $(I_z\ Z)$ from equations (3.2) and (3.4), respectively, into equation (3.6) yields:

$$p_a = (\pi D L B_{av})(\pi D a c)n \times 10^{-3} \quad \text{kW} \tag{3.7}$$

$$= (\pi^2 B_{av} ac \times 10^{-3})D^2 Ln \tag{3.8}$$

$$= C_o D^2 Ln \tag{3.9}$$

where,

$$C_o = \pi^2 B_{av} ac \times 10^{-3} \tag{3.10}$$

and is known as the output coefficient of a d.c. machine.

Equation (3.9) is the output equation of a d.c. machine. It should be noted that P_a is the power developed by the armature, and it must not be confused with the rating of the machine. As the iron and mechanical losses are relatively small in large machines, they may be neglected at the initial calculation stage. However, distinction must be made between motor and generator action. In small machines, the mechanical and iron losses are relatively large and a good starting point is to assume that these losses constitute approximately 30% of the total losses at full load.

3.3.2 Output equation of a.c. machines

Assuming an m-phase machine having one circuit per phase, the kVA developed by its armature, Q, can be obtained from the armature induced emf, E_{ph}, and the phase current, I_{ph}, as follows [3.1, p. 69]:

$$Q = m(\sqrt{2}\pi f\phi T_{ph}K_w)(I_{ph}) \times 10^{-3} \quad \text{kVA} \tag{3.11}$$

Noting that: $f = pn_s/2$, $I_z = I_{ph}$ (as there is only one circuit per phase), and $Z = 2mT_{ph}$, equation (3.11) may be rearranged as follows:

$$Q = \frac{\pi}{2\sqrt{2}}(p\phi)(I_zZ)n_sK_w \times 10^{-3} \tag{3.12}$$

Substituting values of $(p\phi)$ *and* (I_zZ) from equations (3.2) and (3.4) into equation (3.12), we get:

$$Q = \frac{\pi}{2\sqrt{2}}(\pi DLB_{av})(\pi Dac)n_sK_w \times 10^{-3} \tag{3.13}$$

$$= (1.11\pi^2 B_{av}acK_w \times 10^{-3})D^2Ln_s \tag{3.14}$$

$$= C_oD^2Ln_s \tag{3.15}$$

where,

$$C_o = 1.11\pi^2 B_{av}acK_w \times 10^{-3} \tag{3.16}$$

Equation (3.15) is the output equation of an a.c. machine, and C_o is known as the output coefficient.

3.4 FACTORS AFFECTING SIZE OF ROTATING MACHINES

In general, the cost of a machine increases with its size. However, it is sometimes cheaper to utilise available components of an oversize machine rather than to build a machine of an "optimised" design requiring fresh tooling. In fact, in standardised ranges, it is not uncommon to find that a range of outputs is covered by the same frame size.

It can be seen that the output coefficient of rotating machines, C_o, is equal to:

$$\frac{\text{kW}}{D^2Ln} \text{ (d.c. machine)}, \qquad \frac{\text{kVA}}{D^2Ln_s} \text{ (a.c. machine)} \tag{3.17}$$

The volume of the active parts of the machine is closely related to D^2L, and the rated torque is proportional to kW/n or kVA/n_s. Therefore, the output coefficient is proportional to rated torque per unit volume.

The volume of the active parts of a machine, its size and, generally, its cost, decrease with increase in speed and/or increase in the value of the output coefficient. Since the output coefficient is proportional to the product of specific magnetic and specific electric loadings, it is desirable to use the highest possible values of specific loadings in the design. However, using high values of specific loadings could have an adverse effect on the machine performance characteristics such as: efficiency, temperature rise, power factor (in case of induction motors), and commutation (in case of d.c. machines). Therefore, since size and cost are not the only important aspects of machine design, the designer is required to analyse the effect of increased loadings on the overall performance. Some of the factors that affect the performance of rotating electrical machines and need to be considered when choosing rotor speed and specific loadings are discussed below. Only the factors which apply to all types of machines are discussed in this chapter. The factors which are applicable only to particular types are discussed later in this book.

3.4.1 Speed

As the volume of the active material in a machine varies inversely with its rated speed, the highest practical speed rating should be selected. The maximum speed is, obviously, limited by mechanical stresses in the rotor. This, however, is not the only design criterion that needs to be satisfied.

The synchronous speed of an a.c. machine is determined by the number of poles and supply (or line) frequency. In d.c. machines, although the speed is independent of the number of poles, the frequency of the armature current is determined by the number of poles and rotor speed. For a given number of poles, the armature frequency varies directly with rotor speed, and the armature iron losses increase with frequency. The losses, and their distribution within a machine, are among the factors which determine the temperature rise and efficiency. These factors are discussed later in detail; in the sections dealing with the design of each group of machines. Here, it suffices to emphasise that mechanical considerations are not the only factors to be considered when choosing the rated speed of a machine.

3.4.2 Specific magnetic loading

The specific magnetic loading, B_{av}, affects: the maximum flux density in the iron parts, magnetising current, and core losses.

The maximum flux density

The maximum flux density in the iron anywhere in the machine must not exceed the saturation flux density of the material used. In a well designed machine the maximum flux density occurs in the narrowest section of the tooth. If the leakage flux is neglected, i.e. all the flux lines crossing the air-gap are assumed to pass through the teeth, the relationship between the specific magnetic loading, B_{av}, and the teeth flux density, B_t, can be found by equating the total flux at either side of the air-gap, therefore:

$$B_{av}\pi DL = B_t W_t LS, \quad \frac{B_t}{B_{av}} = \frac{\pi D}{W_t S} \tag{3.18}$$

where: W_t is the smallest tooth width, and S is the number of teeth (or slots).

In machines using rectangular parallel-sided slots, and tapered teeth, the maximum flux density occurs at the root of the rotor tooth (in the case of d.c. machines), and at a section near the air-gap of the stator tooth (in a.c. machines). This effect is more pronounced in medium and some small machines (due to their small diameter), and consequently, the ratio of teeth flux density, B_t, to specific magnetic loading, B_{av}, can be very large at the section where the teeth have the smallest width. Therefore, these machines are designed with lower specific magnetic loading than large machines. The ratio of B_t/B_{av} is generally between 2.5 and 3.5 in standard 50-Hz machines.

The magnetising current

The magnetising current of a machine is directly proportional to the mmf (magneto motive force) required to produce the no load flux in the air-gap and the iron parts. The mmf required to pass the flux in the iron sections of a machine is negligible compared with that of the air-gap; provided that the iron is not saturated. The air-gap mmf is directly proportional the specific magnetic loading. Therefore, it can be assumed, in the first instant, that the magnetising current is proportional to the specific magnetic loading.

The value of the magnetising current does not usually present a serious design constraint in d.c. and synchronous machines. This is because these machines, generally, have two electrical inputs; one for the field and the other for the armature. This enables independent control of the field winding design parameters and current. Induction machines, on the other hand, have their magnetising current supplied through the armature winding. Increased value of magnetising current results in a lower operating power factor. Therefore, specific magnetic loading values in the case of induction machines are generally lower than those used for d.c. and synchronous machines.

The core losses

The core (or iron) losses may be considered to consist of two components; the eddy current and the hysteresis loss. The hysteresis loss varies directly with frequency and the square of the flux density. The eddy current loss is, approximately, proportional to the flux density and the square of the frequency. The flux density in the iron parts of a machine is directly proportional to the specific magnetic loading. Therefore, the limiting factor in the choice of specific magnetic loadings for high speed d.c. machines and high frequency a.c. machines is more likely to be the core losses than the saturation level of the lamination material. An obvious example is the induction motors used on aircraft. As their supply frequency is 400 Hz, and in order to keep the core losses at an acceptable level, the specific magnetic loading for such machines would be about half, or even less, the value used for an equivalent machine designed to operate from supply frequency. It is worth noting that such high frequency machines usually utilise more expensive lamination materials, exhibiting lower specific iron losses than standard silicon–iron laminations [3.2].

3.4.3 Specific electric loading

As is the case with specific magnetic loading, it is desirable to use the highest possible specific electric loading. The main limiting factor in the choice of specific electric loading is the temperature rise of the machine.

The copper loss in the machine is proportional to the squared value of the current density in its conductors. The choice of conductor current density is therefore limited by the permissible temperature rise. The maximum allowable temperature rise of a machine is determined by the type of insulating materials used. For example, class A organic materials, such as cotton, can withstand a temperature of up to about 105 °C, while class H inorganic materials, such as polyester films, may operate at higher temperatures (about 180 °C). Obviously, the use of high grade insulating materials in a machine, combined with improved cooling, would allow an increase of its specific electric loading.

Increasing the number of conductors, Z, in a machine necessitates increasing the slot area so that all the conductors may be accommodated. The slot area can be increased by using deep slots, wide slots, or a combination of both. Using deep slots results in increased slot leakage inductance and higher tooth mmf, and, with constant core depth, larger outside diameter. On the other hand, using wide slots results in narrow teeth, for a given frame size. This reduces slot leakage reactance but increases the teeth flux density-to-specific magnetic loading ratio (B_t/B_{av}, equation (3.18) above), and, in extreme cases, could necessitate using a lower value of B_{av} in order to avoid saturating the teeth.

Another factor which needs to be considered when choosing the specific electric loading, although not usually applicable to small machines, is the operating voltage of the machine. The insulating materials in a slot occupy more space in a high voltage machine (several kilovolts) than in a low voltage machine. This results in the slot space factor (ratio between slot space occupied by active conductor material and total slot area) of a high voltage machine being less than that of a low voltage one. Therefore, the specific electric loading of a high voltage machine would, generally, be less than that of a lower voltage one of the same rating.

3.4.4 Discussion

It is shown above that the specific magnetic loading, B_{av}, determines the magnetisation requirements of the machine. A high value of B_{av}, as well as increasing the magnetising mmf and core losses, necessitates the use of wide teeth. Within a given frame size, this will be at the expense of narrow slots which will be able to accommodate a fewer number of conductors; thereby, resulting in a reduction in the specific electric loading, *ac*. The specific loadings compete for space within a given frame size. An increase in one, generally, results in a reduction in the other, and this gives rise to the comparison sometimes made between a *copper* design and an *iron* design.

In a copper design the priority is given to housing a large number of conductors in large slots at the expense of narrow teeth. Therefore, such a design is based around a relatively high value of specific electric loading, *ac*, and a relatively low value of specific magnetic loading, B_{av}. In an iron design the opposite applies.

There are no definitive rules to help the designer to decide whether to go for one design or the other. Instead, the designer has common sense and experience. For example, a 400 Hz aircraft fuel pump motor will be immersed in kerosene during normal operation, and this enables a very much higher current density than normal to be utilised (say 20 A/mm^2 compared with 3–6 A/mm^2 for a standard motor). In view of the high operating frequency, the designer will be aiming to use a low value of B_{av} in order to limit the core losses to an acceptable value. Therefore, in this particular case, there are no conflicting requirements and a copper design will be adopted. If, on the other hand, the design problem is concerned with an induction motor to operate on an oil rig in a tropical country, a relatively low value of specific electric loading will be used, and such a design will be nearer to an iron one. All things being equal, in view of the cost of copper, an iron design would, probably, be cheaper than a copper design.

3.5 VARIATION OF OUTPUT AND LOSSES WITH DIMENSIONS

In order to reduce the number of variants, let us assume constant the number of slots and poles, rated speed and flux and current densities. In other words, the machine geometry is fixed and its size is varied by changing the linear dimensions (D and L) by the same factor.

It should be noted that it is unrealistic to keep the current density constant as the size is varied, and this constraint will be relaxed later. Also, in practice the windings will be redesigned as the size is varied. In this discussion the number of conductors is fixed and the effect of scaling the machine will be taken as scaling the conductor's diameter and length, in direct proportion to D and L respectively.

Because the flux densities in the iron parts of the machine are assumed constant as the size is varied, the specific magnetic loading will not be affected by scaling.

The conductor current may be expressed as the product of the current density, δ, and the conductor cross-section area, a_z. Thus, equation (3.4) may be rewritten as follows:

$$ac = \frac{(\delta a_z)Z}{\pi D} \tag{3.19}$$

Since the conductor cross-section area, a_z, is proportional to D^2, the specific electric loading (and the output coefficient) will be proportional to D. Therefore, for fixed densities, the bigger the diameter the better is the material utilisation in a machine.

Consider two machines of the same type, and let all the linear dimensions of the first (machine A) be x times those of the second (machine B). As D^2 is proportional to x^2, L is proportional to x and the specific electrical loading (ac) is proportional to x, the output of machine A would be x^4 times the output of machine B.

The copper (I^2R) losses can be calculated by multiplying the number of conductors, Z, by the copper loss in each conductor:

$$I^2R = Z \times (I_z)^2 \rho \frac{L}{a_z} = (\delta^2 \rho)(Z a_z L) \tag{3.20}$$

where ρ is the resistivity. Therefore, the copper losses are proportional to the volume of the conductors. Thus, the copper loss of machine A would be x^3 times that of machine B. The specific iron loss (loss per unit volume) remains constant as the flux densities are assumed constant. Therefore, the total losses vary as the third power of linear dimensions, and the total losses of machine A would be x^3 times the losses of machine B.

The heat produced by the losses in a machine has to be conveyed to the surrounding media. The effect of scale on temperature rise due to the iron losses may be disregarded at this stage. This is because the iron is usually in direct contact with the cooling media and the surface area through which heat is transferred will vary directly with the scaling factor.

The thermal resistance between the heat sources in slots and the cooling media, R_{th}, is mainly due to the air and slot insulation [3.3]. If the thickness of insulation is assumed constant, then R_{th} will decrease quadratically with scale, or:

$$R_{\text{th}} \propto \frac{1}{x^2} \tag{3.21}$$

The temperature rise, $\Delta\theta$, is obtained by multiplying R_{th} with the losses, therefore:

$$\Delta\theta \propto \frac{x^3}{x^2} \propto x \tag{3.22}$$

and the temperature rise of machine A would be x times that of machine B. However, the variation of temperature rise with linear dimensions is not strictly correct. This is because as the rotor diameter, and peripheral speed, increase, the ventilation conditions in the machine improve. Furthermore, the cooling systems employed in large machines are more sophisticated than those used in small machines. In fact, electric machines are designed to operate over a very narrow range of temperature rise. Therefore, a more realistic analysis should assume a fixed value of temperature rise.

Keeping the temperature rise constant as the scale is varied results in the current density, δ, being inversely proportional to the square root of the scaling factor x. This explains why the current density is greatest in small machines, diminishing considerably with increase in size. However, in very large machines where direct cooling of conductors is provided, the value of current density can be appreciably increased.

This more realistic assumption regarding the current density makes the specific electric loading, and the output coefficient, proportional to the square root of the diameter (or the scaling factor x); which is still an improvement in utilisation as the scale is increased.

For a given temperature rise, it is estimated that the output and losses of machine A will be $x^{3.5}$ and x^2 times the corresponding values of machine B, respectively. Neglecting the mechanical losses, the efficiency, η, can be calculated by:

$$\eta = \frac{\text{output}}{\text{output} + \text{losses}} \propto \frac{x^3\sqrt{x}}{x^3\sqrt{x} + Kx^2} \propto \frac{1}{1 + \dfrac{K}{x\sqrt{x}}} \tag{3.23}$$

where K is a constant. Equation (3.23) shows that the efficiency increases with increase in linear dimensions. This explains, in part, why sub kilowatt motors have efficiencies of the order of 60% while large alternators have efficiencies over 97%.

3.6 SEPARATION OF *D* AND *L*

The value of D^2L can be calculated using equations (3.9) and (3.15) for d.c. and a.c. machines, respectively. The next step in the calculation is to split this product up into its components D and L.

It is shown above that both the specific electric loading and the output coefficient are proportional to the square root of the air-gap diameter, D. Therefore, in most cases, significant advantages accrue from designing for large diameter and small core length. It should be noted that as the diameter increases, the pole pitch (and the length of the end connections) increases. There is obviously a limit to the L/D ratio below which the end region resistance, losses, flux and cooling requirement offset the advantage of designing for large diameter.

Some of the factors which influence the relative values of D and L are common to all types of machines and some are only relevant to specific types. For example, the rotor peripheral speed and moment of inertia increase as the diameter is increased, and this applies to all types of machines. On the other hand, the maximum core length of a d.c. machine, for example, is determined by the maximum allowable voltage between two adjacent commutator segments.

A peripheral speed of about 30 m/s does not call for any special rotor construction. Fortunately for the designers of small machines, the peripheral speed is unlikely to exceed this value. The rotor moment of inertia is approximately proportional to D^4L. Therefore, the diameter of machines designed for control systems applications should be made as small as is practically feasible. Conversely, high inertia machines may be required for impact-load applications, and these should be designed for large diameters.

3.6.1 Separation of *D* and *L* for d.c. machines

The pole cross-section requiring the smallest volume of copper for the field winding is circular. However, circular poles necessitates the use of solid pole sections, which is not the current practice. This is mainly because the use of laminated poles yields reduction of production cost.

When using laminated poles, the length of the field winding mean turn is minimum when the periphery is in the form of a square. This means the core length L should be chosen to be approximately equal to the pole arc. In practice, the core length is usually made slightly greater than the pole arc. Since the pole arc/pole pitch ratio is normally between 0.65 to 0.75, L/τ is usually between 0.6 and 0.9. In most cases, small d.c. machines should be designed for large diameter rather than large core length.

The maximum allowable voltage between adjacent commutator segments is, typically, 30 V. Assuming a rotor (armature) peripheral speed of 30 m/s, maximum air-gap flux density under load conditions of 1.0 T and single turn armature coils, the maximum core length would be approximately 0.5 m. This should not inconvenience the designer of small d.c. machines in any way. However, this limitation together with the limitation on the maximum commutator diameter (which is approximately 0.75 the rotor diameter) determine a maximum size for d.c. machines. This explains why d.c. machines are not found with power ratings of more than 5 MW.

3.6.2 Separation of *D* and *L* for synchronous machines

The diameter of large alternators is limited by the maximum allowable peripheral speed. Small synchronous machines may be regarded as line-start permanent magnet motors and small alternators. Small alternators, with ratings of between 5 and 30 kVA, may be used to provide emergency power supplies and as permanent supplies to stationary plants not connected to the mains, or to mobile installations. Apart from aircraft alternators, these machines operate at a maximum speed of 3600 rev/min.

The ratings and operating speeds of small synchronous machines do not normally require special rotor construction or special cooling. The diameter is therefore determined by the number of poles; the designer needs to ensure that the pole pitch is sufficiently large so that the field coils can be accommodated. As with d.c. machines, the volume of copper is minimum when the field coils cross-section is approximately a square. As far as small synchronous machines are concerned, a core length/pole pitch ratio (L/τ) of between 0.8 and 1.2 yields satisfactory designs in most cases.

3.6.3 Separation of *D* and *L* for induction motors

While aiming to design for large diameters to realise the benefit of increased output coefficient, the designer of induction motors needs to ensure that the machine operates with a satisfactory power factor. The magnetising current of an induction machine decreases (and the power factor improves) as the diameter is increased, but it increases as the number of poles is increased. A motor having large pole pitch ($\tau = \pi D/p$) and small number of poles will have a higher power factor than a motor with a small pole pitch. L/τ ratio between 1 and 1.25 yields a good power factor, and a value between 1.5 and 2 gives minimum cost. The diameters and lengths of motors with a high value of pole pitch are selected to give minimum cost, while these values are proportioned to give a good power factor at reasonable cost for motors with low value of pole pitch.

It should be noted that it is not always possible to use L/τ ratio greater than or equal to unity for fractional-horsepower motors and some 2-pole motors below, say, 11 kW rating (at line frequency). This is because the resulting small diameter will necessitate the use of too small a number of slots. L/τ value as small as 0.6 is not uncommon in fractional-horsepower motors.

3.7 STANDARD FRAMES

Apart from special machines, the manufacture of motors for industrial applications is concentrated into a series of standard frames covering a wide range of power ratings. A *frame* is the mechanical structure housing a stator of a given outside diameter along with the bearings, cooling fan (if used), end covers and terminal box.

All small and medium size motors are built with standard frame sizes as specified in *IEC 72*, 1971, and its amendments (1977 and 1981). This publication gives a comprehensive range of main structural dimensions of foot-mounted motors with centre heights between 56 and 400 mm, and flange-mounted motors with flange diameters between 55 and 1080 mm.

Although the object of standardisation is to enable the interchangeability of machines made by different manufacturers all over the world, this is limited because of the difference in nationally standardised voltages. For example, while the industrial three-phase low-level supply voltage in Britain, Australia and New Zealand is 415/240 V, 50 Hz, the continental standard is 380/220 V, 50 Hz, and the North American practice is 208/120 V, 60 Hz. 660/380 V, 50 Hz is the practice in Finland. India and Pakistan's system is 400/230 V, 50 Hz, and Mexico adopts a 220/127 V, 60 Hz system. The majority of ships and offshore installations operate at 60 Hz with a mixture of the forementioned voltage levels.

3.7.1 Designation of machines

For foot-mounted machines, the machine's centre height, expressed in millimetres, designates the frame size, or the frame number. When more than one core length

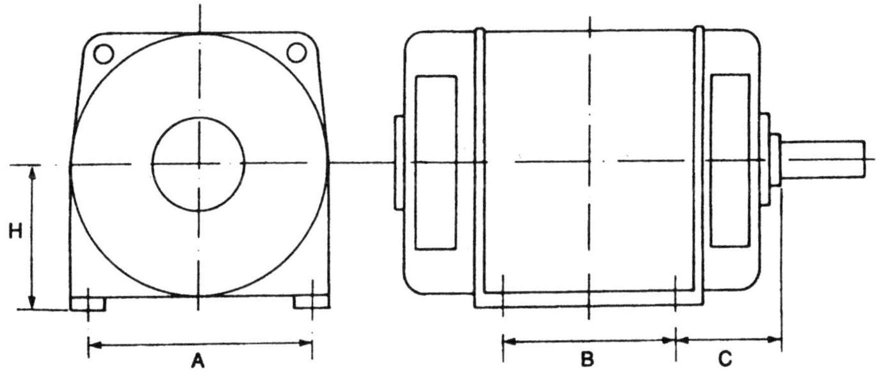

Figure 3.1 Standard symbols for the principal fixing dimensions of horizontal-shaft motors.

is used, the frame number will consist of the machine's centre height followed by a letter indicating the core length; letters S, M and L indicate short, medium and long core, respectively.

Variation of the core length allows two or more ratings to be provided by the same frame. For example, a 1.5 kW, 4-pole motor would have an air-gap diameter of 80 mm and a core length of 100 mm ($L/\tau = 1.59$). Reducing the core length of this machine to 75 mm and utilising the same laminations, an output of about 1.1 kW will be obtained. The height of the shaft centre of both machines will be the same; frames 90L (long) and 90S (short) will be used for the 1.5 kW and 1.1 kW, respectively.

A letter appearing before the frame number indicates the type of enclosure: C indicates enclosed ventilated, D indicates totally enclosed and E indicates flameproof.

The diameter of the shaft, in millimetres, follows the frame number. A dash is used to separate the frame number and the shaft diameter if the frame number does not end with a letter. For example, D71-20 denotes a totally enclosed machine (D) whose centre height is 71 mm and shaft diameter is 20 mm.

When a foot-mounted machine is also provided with a flange at the driving end, the flange number may be added after the shaft diameter. Machines having only flange mounting are designated by the shaft diameter followed by the flange number.

The system of frame nomenclature is internationally agreed up to and including frame size 400. However, the Standards are mainly concerned with a.c. induction motors. The dimensions of d.c. machines are not fully standardised, nor are outputs and shaft numbers assigned to particular frame size.

3.7.2 Dimensional standards

IEC Publication Number 72 gives the full details of standardised dimensions of foot-mounted and flange-mounted machines. With reference to Figure 3.1, the main fixing dimensions of a foot-mounted machine are:

Table 3.1 Fixing dimensions of foot-mounted a.c. machines with centre heights 56–200 mm [3.4]. (All dimensions in millimetres)

Frame no.	H	A	B	C
56	56	90	71	36
63	63	100	80	40
71	71	112	90	45
80	80	125	100	50
90S	90	140	100	56
90L	90	140	125	56
100S	100	160	112	63
100L	100	160	140	63
112S	112	190	114	70
112M	112	190	140	70
132S	132	216	140	89
132M	132	216	178	89
160S	160	254	178	108
160M	160	254	210	108
160L	160	254	254	108
180S	180	279	203	121
180M	180	279	241	121
180L	180	279	279	121
200S	200	318	228	133
200M	200	318	267	133
200L	200	318	305	133

standardized centre height H (frame number),
distance between centre lines of fixing holes A and B, from end view and side view, respectively, and;
distance from shaft shoulder to centre line of mounting holes at driving end C.

Table 3.1 gives the main standardised fixing dimensions of foot-mounted machines for frame sizes 56 to 200.

3.7.3 Standardised ratings

The IEC recommended ratings in kilowatts of machines covered in this book are: 0.06, 0.09, 0.12, 0.18, 0.25, 0.37, 0.55, 0.75, 1.1, 1.5, 2.2, 3.7, 5.5, 7.5, 11, 15, 18.5, 22, 30, 37.

These 20 different ratings can be provided with only eleven standard frames; utilising short, medium and long stators for some.

Tables 3.2–3.5 relate the selected range of standard ratings to frame sizes, and, they give values of air-gap diameter (D) and core length (L) for standard three-phase motors designed with 2, 4, 6 and 8 poles.

Table 3.2 Frame numbers, outputs and main dimensions of standard 2-pole three-phase motors

Frame no.	kW	*D* (mm)	*L* (mm)
56-15	0.09	40	45
56-15	0.12	40	55
63-18	0.18	45	45
63-18	0.26	45	60
71-20	0.37	55	50
71-20	0.55	55	65
80-25	0.75	62	60
80-25	1.1	62	80
90S30	1.5	70	75
90L30	2.2	70	100
100L36	3.0	80	90
112M38	4.0	90	110
132S50	5.5	110	90
132S50	7.5	110	120
160M55	11	135	130
160M55	15	135	180
160L55	18.5	135	215
180M65	22	150	190
200L75	30	170	200
200L75	37	170	250

Table 3.3 Frame numbers, outputs and main dimensions of standard 4-pole three-phase motors

Frame no.	kW	*D* (mm)	*L* (mm)
56-15	0.06	45	45
56-15	0.09	45	55
63-18	0.12	50	45
63-18	0.18	50	60
71-20	0.25	62	50
71-20	0.37	62	65
80-25	0.55	70	60
80-25	0.75	70	80
90S30	1.1	80	75
90L30	1.5	80	100
100L36	2.2	90	90
100L36	3.0	90	120
112M38	4.0	103	140
132S50	5.5	125	125
132M50	7.5	125	170
160M55	11	150	160
160L55	15	150	215
180M65	18.5	170	200
180L65	22	170	240
200L75	30	190	250

Table 3.4 Frame numbers, outputs and main dimensions of standard 6-pole three-phase motors

Frame no.	kW	*D* (mm)	*L* (mm)
56-15	0.06	50	55
63-18	0.09	55	45
63-18	0.12	55	60
71-20	0.18	70	50
71-20	0.25	70	65
80-25	0.37	80	70
80-25	0.55	80	90
90S30	0.75	90	75
90L30	1.1	90	100
100L36	1.5	103	120
112M38	2.2	115	140
132S50	3	135	125
132M50	4	135	150
132M50	5.5	135	180
160M55	7.5	165	165
160L55	11	165	225
180L65	15	180	240
200L75	18.5	200	230
200L75	22	200	265

Table 3.5 Frame numbers, outputs and main dimensions of standard 8-pole three-phase motors

Frame no.	kW	*D* (mm)	*L* (mm)
63-18	0.06	55	45
63-18	0.09	55	60
71-20	0.12	70	50
71-20	0.18	70	65
80-25	0.25	80	70
80-25	0.37	80	90
90S30	0.45	90	75
90L30	0.55	90	100
100L36	0.75	103	90
100L36	1.1	103	120
112M38	1.5	115	140
132S50	2.2	140	100
132M50	3	140	135
160M55	4	170	125
160M55	5.5	170	165
160L55	7.5	170	225
180L65	11	190	240
200L75	15	215	265

3.8 REFERENCES

[3.1] McPherson, G. and Laramore, R. D. (1990) *An Introduction to Electrical Machines and Transformers*, New York: John Wiley & Sons.

[3.2] Chalmers, B. J. *et al.* (1980) *Proc. IEE, Pt. B*, **6**, p. 341.

[3.3] Hamdi, E. S. (1983) *ERA Report No. 83-0167*, ERA Technology Ltd, Leatherhead, Surrey, England.

[3.4] International Electromechanical Commission (1971) Dimensions and Output Ratings for Rotating Electrical Machines — Frame Numbers 56 to 400 and Flange Numbers F55 to F1080, *IEC Publication No. 72* (5th edn).

4

DESIGN OF DIRECT CURRENT MACHINES

4.1 INTRODUCTION

Although d.c. machines can work as motors or generators, their present use is almost exclusively as motors. The almost universal use of a.c. generation, transmission and distribution systems is due to economic as well as technical reasons. At present the use of d.c. generators is limited to some small standby generating plants driven by windmills and some uninterruptible power supply units. To this end, the treatment of d.c. generators is excluded.

The decline of d.c. generators as a source of electric energy and the development of reliable and affordable static semiconductor converters has affected the popularity of d.c. motors in many industrial applications. In spite of this, d.c. motors still occupy an important place in the field of variable speed drives. Their industrial uses include rolling mills, overhead cranes, some machine tools and other applications requiring precise control of torque and/or speed. They are also used in battery driven vehicles (such as milk floats) and fork trucks and for driving railway locomotives. Small d.c. motors, mostly of the permanent magnet type, are extensively used in automotive vehicles. The starter motors, windscreen wiper motors, window motors, power steering motors, fan motors and motors to drive other accessories are all d.c. motors. Miniature d.c. motors find their place in many hand-held battery-operated tools and appliances.

4.2 CONSTRUCTIONAL DETAILS

The conventional d.c. machine may be regarded as consisting of three identifiable components; a field system, an armature and a commutator. Figure 4.1 shows the basic construction of a d.c. machine. The field system of a d.c. machine is located on its stator, while the armature is housed on the rotor.

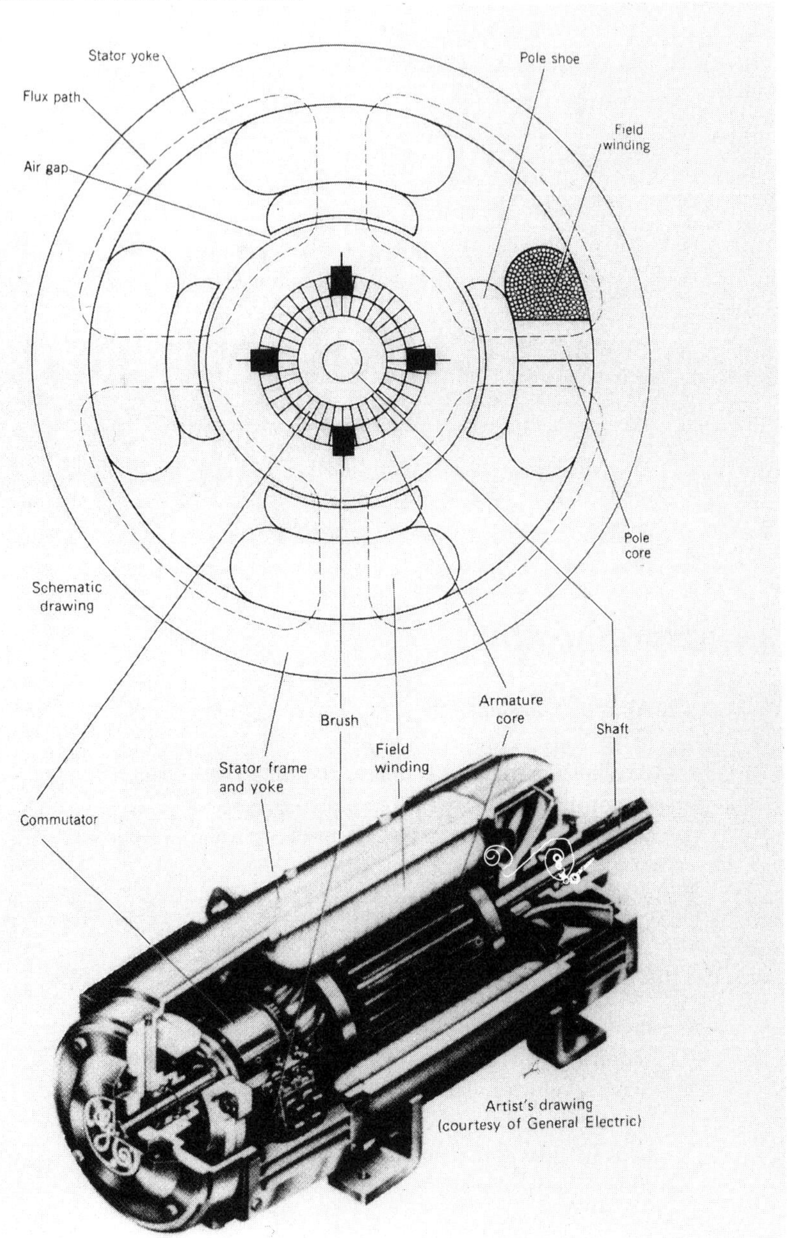

Figure 4.1 Construction of a d.c. machine.

It should be noted that the terms *field* and *armature* refer to electrical circuits which are essential for the process of electromechanical energy conversion to be performed. The *stator* and *rotor* refer to the stationary and the rotating parts of an electromechanical device; respectively. In a conventional (mechanically commutated) d.c. machine the field system is placed on the stator and the armature on the rotor but this is not essential to the concept of energy conversion. The position of the two windings is reversed in some unconventional designs (for example, in electronically commutated machines) and in synchronous machines.

The field system provides the machine's excitation. In the general case, the field system consists of main poles, interpoles (or commutating poles) and compensating windings. The main poles are designed to provide the main steady magnetic field necessary for the operation of the machine. The interpoles, when used, are positioned between the main poles and are designed to improve the commutation conditions in the machine. In case of a machine provided with compensating windings, the pole face (of the main pole) is slotted to accommodate the windings. Compensating windings serve to improve commutation in a similar way as the interpoles. It should be noted that compensating windings are extremely expensive. Their use in small motors is limited to applications involving sudden reversal and high acceleration.

The poles constituting the field system of a d.c. machine are either fitted to or, in the case of some small machines, punched as an integral part of the stator yoke; which provides the return path of the magnetic flux lines.

The electromechanical energy conversion process takes place in the armature. The armature winding of a conventional d.c. machine is housed in longitudinal slots cut in the laminated rotor body. The slots are sometimes skewed in small machines to yield quieter operation and a slight reduction in losses. However, skewing adds to manufacturing cost.

The commutator is always mounted on, but insulated from, the shaft of the d.c. machine at the non-drive end. It consists of copper segments insulated from each other and connected to the coils of the armature winding. Current is conducted to the armature winding by carbon brushes which are held against the commutator surface by the force of springs fitted to the brush-holders. The commutator and brushes perform mechanical rectification (a.c. to d.c.) in case of generators or inversion (d.c. to a.c.) in case of motors.

4.2.1 The stator of a d.c. machine

The stator of a d.c. machine consists of the yoke (or stator core), the main poles and commutating poles.

As well as being a part of the magnetic circuit, the yoke serves as a mechanical support of the field system. Traditionally, solid yokes were used. Solid cast iron yokes were used in earlier machines. These were superseded by cast steel yokes. As cast steel has a saturation flux density value of about twice that of cast-iron, the use of cast steel resulted in significant reduction in the yoke's cross-section and in the weight and size of the machine.

The modern trend is to use laminated yokes made from electrical steels. This is because most modern industrial motors are fed from static power converters which, as well as providing controlled variable supply, introduce harmonic currents in the machine.

Traditionally, the main poles of d.c. machines were cast, either separately or as an integral part with the yoke; and this is still the case in some small machines. In some larger machines the practice was, and still is in some cases, to use laminated pole-shoes attached to solid poles. The modern practice, however, is to use

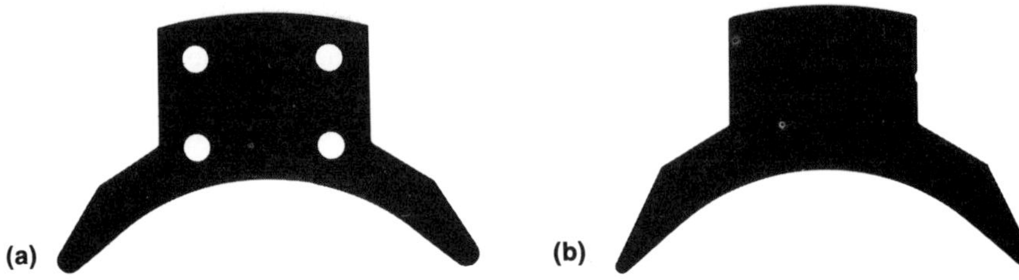

Figure 4.2 (a) Riveted-pole and (b) welded-pole laminations.

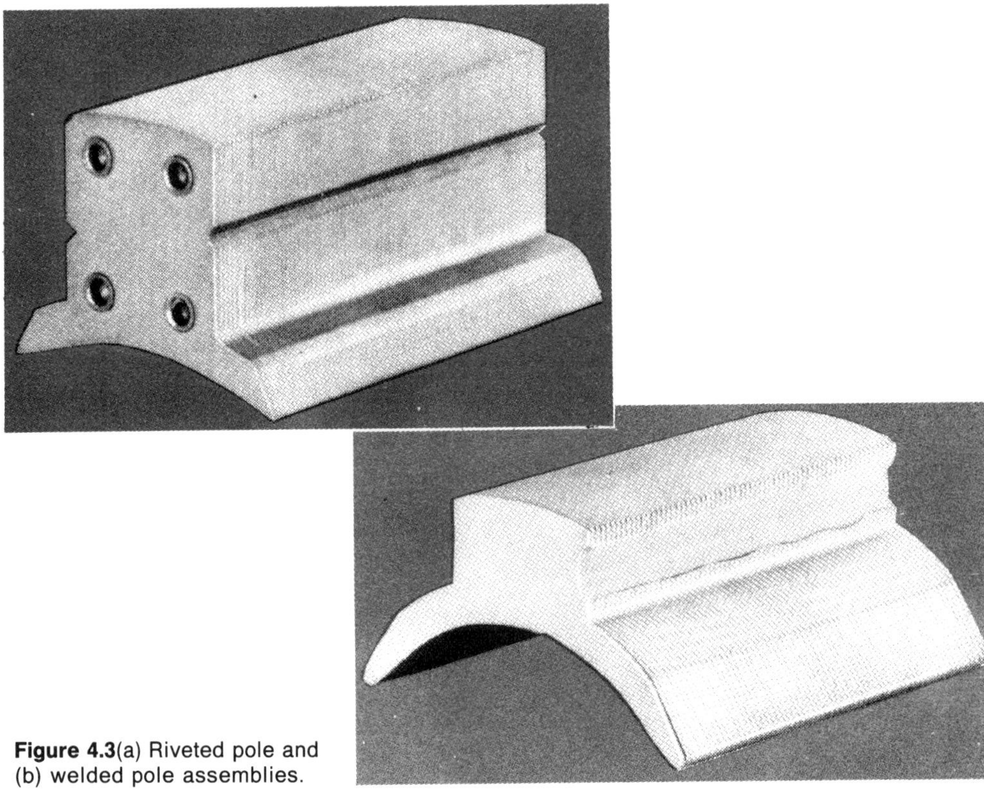

Figure 4.3(a) Riveted pole and (b) welded pole assemblies.

completely laminated poles. The use of laminated poles enables a range of ratings to be realised in the same frame size, but with different core lengths.

The laminations used in completely laminated poles are provided with either rivet holes or welding groves; as illustrated in Figure 4.2. Riveted poles are manufactured by stacking the laminations on steel rods and pressing the whole assembly between steel end plates. The rods are then riveted and the rivet heads are welded. In the case of welded poles, after pressing the laminations the weld is applied directly to the grooves. Figure 4.3 shows a riveted and a welded pole assembly.

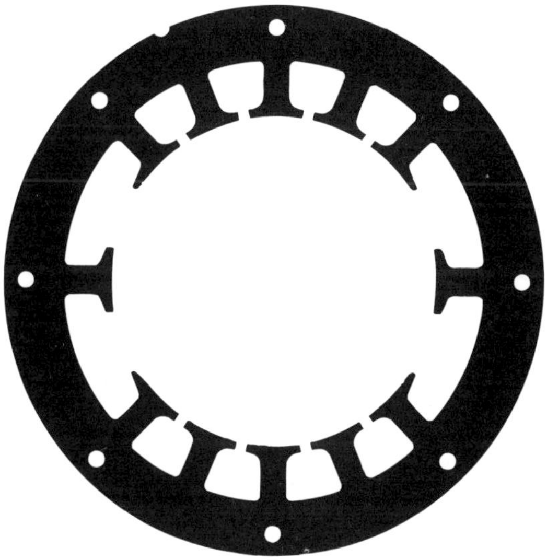

Figure 4.4 A 2-pole stator lamination.

The method most commonly used for attaching poles to yokes in small machines is by drilling and tapping the assembled poles to receive pole bolts.

The minimum length of mean turn of field coil is obtained when the pole cross-section is circular. This can be achieved in the case of cast steel poles. When laminated poles are used, the field coil mean turn length is a minimum when the pole cross-section is square.

The interpoles are made from laminated steel or from low carbon steel. In some small machines solid low carbon steel commutating poles are used. In both cases, parallel-sided interpoles are used.

4.2.2 Stator laminations

The requirements of small d.c. machines can be met by laminated stators manufactured by stacking laminations stamped to provide main pole, interpoles and, in some cases, slots to accommodate the compensation windings. Figures 4.4 and 4.5 show such laminations for 2-pole and 4-pole machines, respectively.

A square lamination providing the yoke, frame, interpoles and main poles is shown in Figure 4.6. Such laminations are suitable for the construction of non-housed machines or machines having square frames. With reference to Figure 4.6, it is shown that lateral slots are provided in the interpoles. This allows insertion of bar sections of different sizes in order to adopt the interpole to the number of armature slots used.

Rivet holes are provided in the laminations shown in Figures 4.4–4.6, which enables the construction of riveted stators as described above.

Up to frame size 180 motors it is generally possible to utilise one-piece laminations to manufacture the stator. For larger sizes, it becomes difficult to wind the main

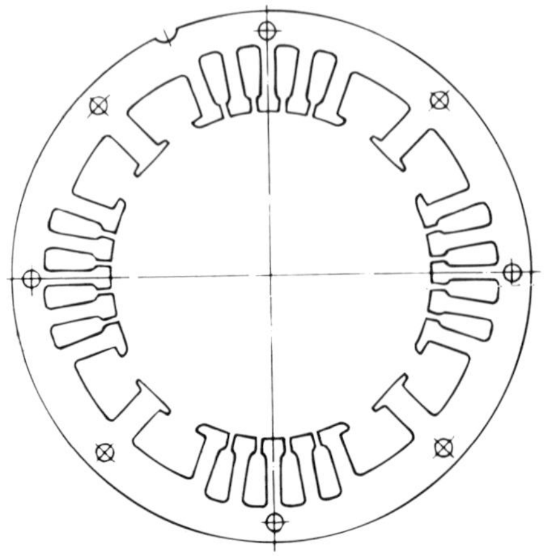

Figure 4.5 A pole stator lamination.

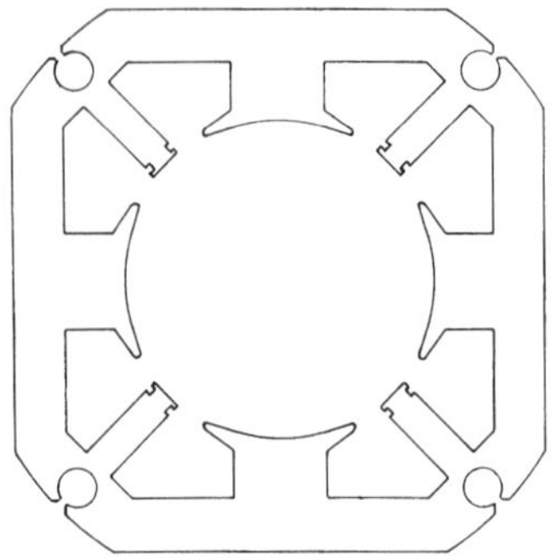

Figure 4.6 A square stator lamination.

field coils on the assembled stator. This necessitates the use of separate laminations for the main pole; the pole is fabricated and the field coil fitted to it before the whole assembly is screwed in place.

4.2.3 Field windings

When the main poles are punched in the stator laminations, the main field coils are wound directly on the poles, with layers of insulation placed between the pole body and winding. When the poles are fabricated from loose laminations, or when cast steel poles are used, the main field coils are wound on suitable formers, insulated and then fitted on to the poles. The form-wound coil arrangement allows repair work to be carried out relatively easily.

Round copper wires are generally used in small d.c. machines. The series field coils may be placed on the top of the shunt coils or they can be arranged as a separate winding. In either case, the two windings must be adequately insulated from each other. When interpoles (commutating poles) are used in a small d.c. machine, the coils are wound directly (of round wire) on the interpole body.

4.2.4 Permanent magnet motors

In a conventional permanent magnet d.c. motor, excitation is provided by an arc-shaped magnet segment fitted directly to the yoke; the magnet assembly replacing the main poles and field windings of a wound-field motor. This generally results in lighter and smaller motors as well as higher efficiency (as a result of eliminating field losses). This construction is very popular in automotive vehicles applications and in powered wheelchairs and forklift drives.

Depending on the permanent magnet material used, the magnet pole pieces are either magnetised in the radial or in the circumferential direction. The low coercive

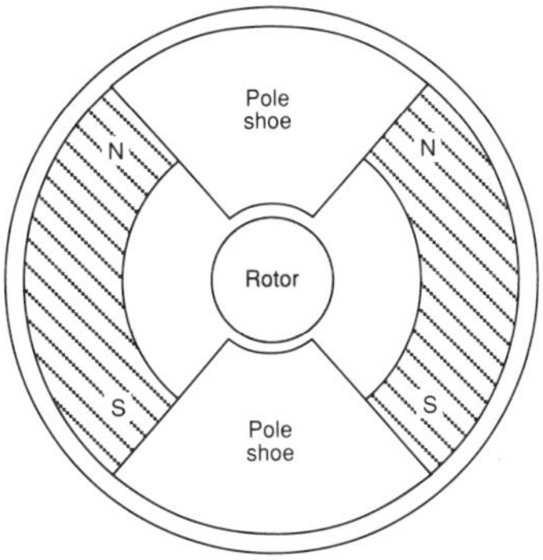

Figure 4.7 Alnico magnet field structure.

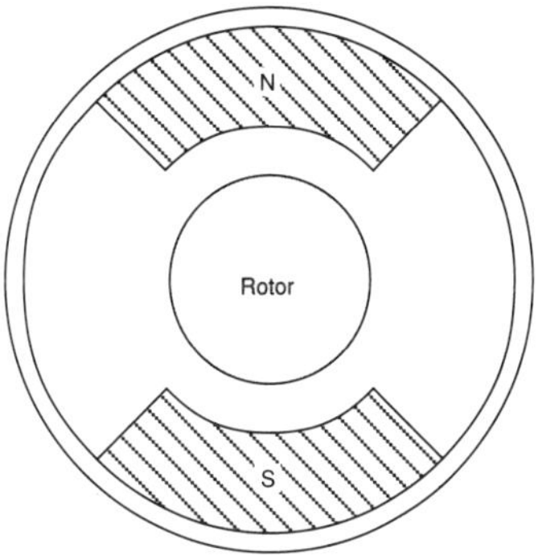

Figure 4.8 Ceramic magnet field structure.

Figure 4.9 A pole ceramic magnet motor.

value of the Alnico magnets necessitates circumferential magnetisation; resulting in an excitation system similar to that shown in Figure 4.7. Additionally, in view of their low value of coercive force, which makes them vulnerable to external demagnetising fields, Alnico magnet poles are usually provided with iron pole shoes to shield the magnets from the demagnetising armature reaction fields.

When ceramic (ferrite) or rare earth magnets are used, their high coercivity permits magnetisation in the radial direction. The resulting structure will be similar to that shown in Figure 4.8; with the rare earth system requiring thinner magnet pieces. Figure 4.9 shows the stator of a 4-pole d.c. motor utilising ceramic magnets. The required axial length of the magnet pole in this particular motor is 165 mm.

In the machine shown in Figure 4.9 there are two segments per pole. The pre-magnetised magnet segments are simply glued to the cast steel yoke with an industrial adhesive.

The field system of larger motors utilising Alnico magnets is usually provided with magnetising windings around the permanent magnet poles. This arrangement permits magnetisation *in situ* after assembly, and demagnetisation to carry out any repair work. Because of the much higher magnetising requirements of ceramic magnets (875–1800 kA/m), samarium–cobalt magnets (1590–3980 kA/m) and neodymium–iron–boron magnets (1590–2790 kA/m), compared with 240–480 kA/m of Alnicos, it is extremely difficult, may be even impossible, to magnetise these magnets *in situ*. Assembling pre magnetised magnet segments adds to manufacturing cost. In any case, it is impractical to install magnetising windings in small ceramic permanent magnet motors, and, the high cost of rare earth magnets is, in most cases, tolerated in order to reduce the size of the motor. On the other hand, pole shoes are not required for ceramic and rare earth magnet motors. Additionally, the recoil permeability of ceramic and rare earth magnets is very close to the permeability of free space. This enables motors utilising these materials to be disassembled and reassembled without changing the magnet operating point, i.e. without affecting the motor's performance.

Although permanent magnet excitation systems may be realised for six or more poles, 2-pole and 4-pole configurations are the most widely used.

In addition to the two configurations shown schematically in Figures 4.7 and 4.8, many other variants exist. For example, with a square frame it is possible to utilise rectangular magnet segments made from rare earth materials. This reduces materials' waste and tooling cost.

4.2.5 Armature

Whereas relatively thick laminations (up to 2 mm thick) or even solid poles (with or without laminated pole shoes) may be used in the fabrication of the field system, only thin (0.35, 0.5, 0.65 mm thick) low loss silicon steel laminations are used in the armature of a d.c. machine. The armature laminations of small d.c. machines are always punched in one piece. They are sometimes provided with axial ventilation holes as illustrated in Figure 4.10.

In small machines, the armature laminations are assembled directly on the shaft and pressed between end flanges. No radial ventilating ducts are used in small machines and axial holes are only used in a small number of cases. Figure 4.11 shows the armature assembly of a small d.c. machine in which no axial ducts are used.

4.2.6 Armature winding

The armature winding of a small d.c. machine utilises multi-turn coils of round copper conductors. The armature coils are placed in the slots with the required insulation between the core and coils. The coils may be held in position in the slots

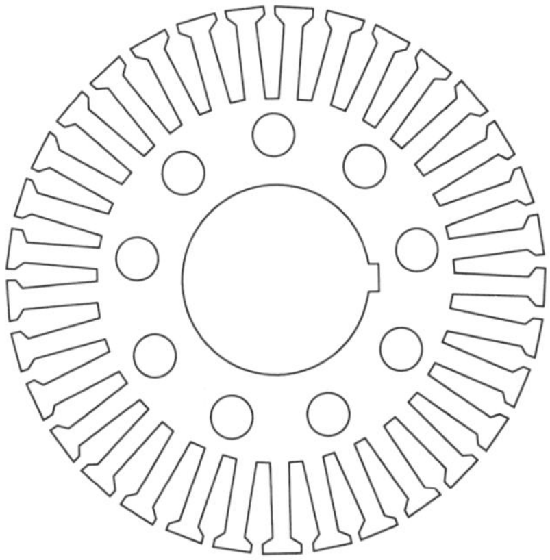

Figure 4.10 An armature lamination provided with ventilating holes.

Figure 4.11 A small d.c. motor armature.

by bands of steel wire wound under tension around the core and/or by slot wedges. Figure 4.12 shows a partially wound armature. In some cases the overhang is held in position by polyester or cotton banding tapes.

The armature winding of a d.c. machine is always a multilayer one; and in the case of small machines it is a double layer one. Simplex lap or simplex wave windings may be used. However, wave winding is preferred as there can be no voltage unbalance between paths in this type of winding (the voltage induced in each path is due to all the poles in the machine). Generally, the choice between the two types of windings is mainly governed by the current per path of the armature winding which has a maximum practical value of 250 A; limiting the armature current of a wave-wound machine to 500 A. This constraint should not really trouble the designer of small d.c. motors and wave windings are almost always used in small machines even in those having more than two poles.

Figure 4.12 A partially wound d.c. machine.

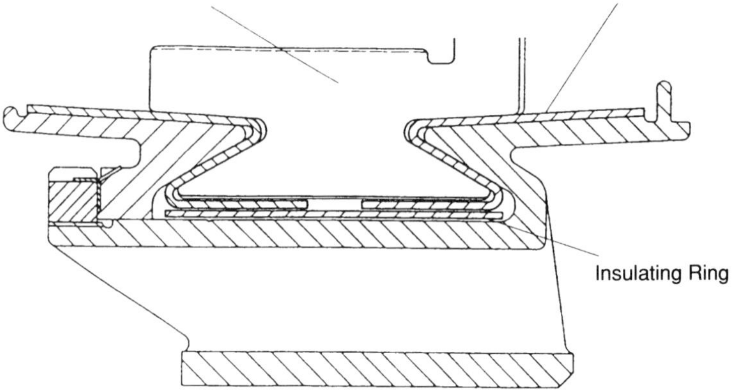

Figure 4.13 Two V-ring commutator construction.

4.2.7 Commutator

The commutator of a small d.c. machine consists of a number of copper segments insulated from one another and connected to the armature conductors. The commutator segments are made from hard-drawn copper or silvered-copper (containing about 5% silver) and they are provided with tows to which armature conductors are soldered. The segments are usually wedge shaped as illustrated in Figure 4.13. The advantage of silvered-copper over hard drawn copper is that it

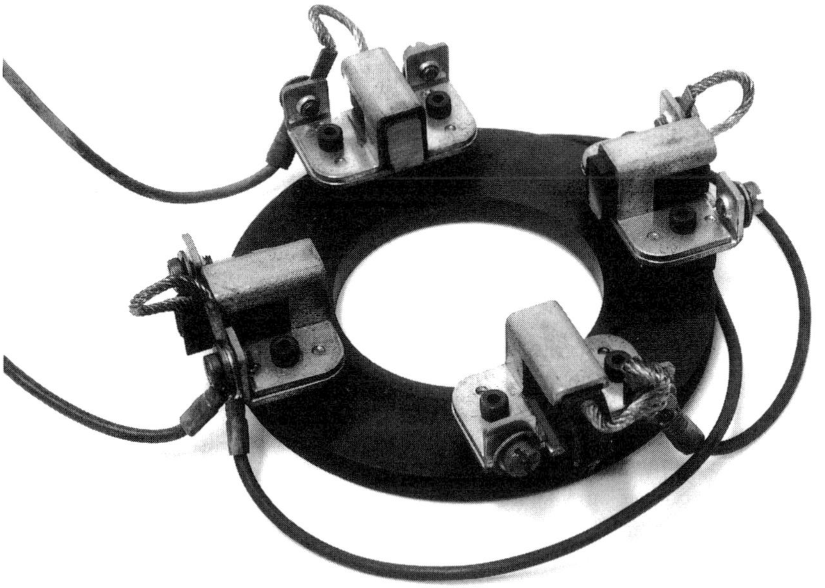

Figure 4.14 Brush gear of a small d.c. motor.

can withstand the flood soldering of the armature coil ends to the tows at 300 °C.

The separators used for insulating commutator segments from each other should have consistent mechanical properties. This is because the quality of commutation depends, among other factors, on accurate segment-to-segment spacing; which obviously depends on the thickness of the insulation. Additionally, the hardness of spacers used for commutator insulation should be comparable to that of copper; so that the segment and insulation wear at approximately the same rate. Spacers made from shellac-bound mica splitting (96% mica) or from epoxy-resin-bound fine mica (the so called Samicanite) are commonly used. The grade of mica used and the type of binder determine the hardness and elasticity of the spacer.

The insulation and copper segments are clamped between two steel V-shaped rings and insulated from them by mica splitting bonded with shellac (or bonded with shellac and epoxy). The two V-ring construction commonly used is illustrated in Figure 4.13. The assembled commutator is then pressed on to the shaft of the machine.

4.2.8 Brush gear

The brush gear consists of a set of brush-holders which are fitted on a yoke of suitable insulating material and house carbon brushes of suitable conductivity and hardness; as illustrated in Figure 4.14.

A brush-holder is simply a metal box open at the top and bottom, provided with an adjustable spring and is mounted close to the commutator surface. The brush is inserted from the top end of the holder and pressed against the commutator surface at the other end by a spring.

The manufacture of brushes having the proper electrical conductivity and sliding-friction characteristics is a very specialised scientific art. Brushes made of natural graphite, hard carbon, electro-graphite or metal graphite may be used.

Natural graphite used with a resin binder yields brush material having good mechanical characteristics at high speed but at the expense of low conductivity (due to the resin content). The low conductivity results in a high contact voltage drop which helps the commutation process. Therefore, brushes made of natural graphite are suitable for use in small motors which do not usually employ interpoles.

Hard carbon yields cheap but hard brushes. Their hardness results in a somewhat high rate of commutator wear. However, the rate of wear is almost the same for commutator segments and insulating separators. These brushes are mostly used in fractional-horsepower machines and in some larger low speed motors where the commutation conditions are not too severe.

Electro-graphite brushes have a low rate of wear and, therefore, they are widely used in industrial motors and in traction applications.

Metal graphite brushes have high conductivity which results in a low value of voltage contact drop. This type of brush is suitable for low voltage high current applications such as automobile starter motors.

Current is carried to the brush by a flexible conductor called a *pigtail*, which may be soldered to the brush or attached by a screw; soldering is mostly used in brushes employed in small motors.

There are three common ways of applying brushes to the commutator. These are: radial, trailing and reaction. In most cases the brushes are applied radially, i.e. their centre line is radial to the commutator. This method results in a motor with similar performances in both directions of rotation. In the radial arrangement the force acting along the direction of rotation tends to cause a tilting action which could result in side wear of the brush. To minimise this, the brushes should be closely fitted in deep brush boxes. The brush height is always greater than either of its cross-sectional dimensions.

In non-reversing high speed machines the brushes may be inclined either in the trailing or in the leading direction; the brush box is given either a trailing or a reaction rake; as illustrated in Figure 4.15. In the trailing box arrangement the vertical angle of inclination is about 10–15°. In this case, the spring, friction and radial force combine to maintain contact of the brush with the left face of the box; thereby improving stability.

In the reaction box arrangement, the brushes are inclined in the leading direction by an angle between 30–40° from the vertical. In this case, the contact force between the right hand side of the brush and the brush-holder side opposes the friction force. This arrangement is preferred when narrow brushes are used.

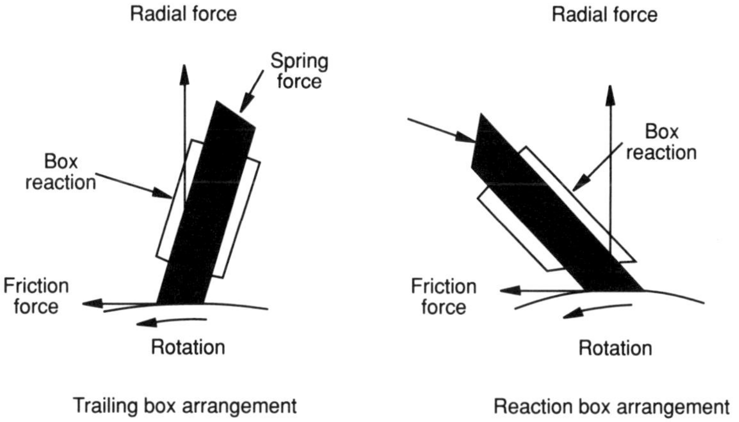

Figure 4.15 Positioning of brushes.

4.3 DESIGN CALCULATIONS

The output equation of a d.c. machine is derived in Chapter 3 and is reproduced here for convenience:

$$P_a = C_o D^2 L n \tag{4.1}$$

and,

$$C_o = \pi^2 B_{av} ac \times 10^{-3} \tag{4.2}$$

where P_a is the armature power (EI_a), kW, B_{av} is the specific magnetic loading or the average flux density in the air-gap, T, ac is the specific electric loading, ampere-conductor/m, D and $L$are the air-gap diameter and core length, m; respectively and n is the speed, rev/s.

Distinction must be made between the armature power (P_a) as used in equation (4.1) and the power rating of the machine. P_a exceeds the power rating of a generator by the armature and series field copper losses and the brush contact losses. For a motor, P_a exceeds the power rating by the core and mechanical losses. The armature power may be assumed equal to the rated power of a motor in the initial design stage. However, it should be noted that in high speed motors the core and mechanical losses are of an appreciable value.

For some applications, for example, automobile starter motors, motors are designed to meet torque rather than power requirement. To enable determination of the diameter and length of a motor required to develop a given stalled torque, equation (4.1) is divided by the angular rotor speed; therefore,

$$T = \left(\frac{\pi}{2} B_{av} ac\right) D^2 L \tag{4.3}$$

where T is the stalled torque, Nm, and ac is the specific electric loading value

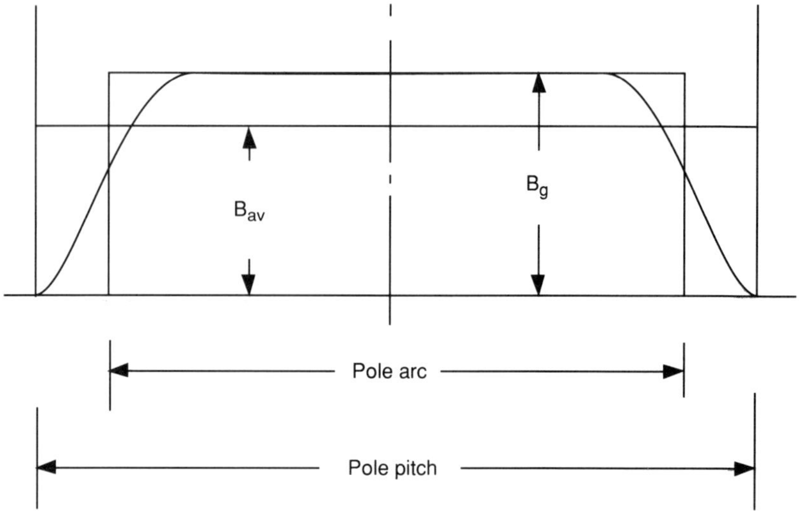

Figure 4.16 Actual and idealised pole flux density.

corresponding to the stall current.

By defining the average air-gap flux density (B_{av}), we assumed that the flux density over the entire pole pitch is constant; equal to B_{av}. In fact, the actual flux density distribution over the pole pitch increases from zero at the interpolar region to a maximum value, B_g, under the pole arc as illustrated in Figure 4.16. The ratio under the true flux distribution curve to the area under the assumed rectangular distribution is called the field distribution factor, K_f. K_f may be taken,approximately, as the ratio of the pole-arc to the pole pitch; therefore,

$$B_g = \frac{B_{av}}{\psi}, \qquad \psi = \frac{\text{Pole arc}}{\text{Pole pitch}} \tag{4.4}$$

The value of ψ is usually between 0.6 and 0.75.

4.3.1 Air-gap density

The main, but not the only, limiting factor in choosing a value for the air-gap flux density in a d.c. machine is armature teeth and core saturation. It is possible to use parallel-sided teeth in most small motors. In the general case, however, the maximum tooth density occurs at the root of the tooth and is calculated as follows:

$$B_{tr} = \frac{B_g \pi D}{w_{tr} S} \tag{4.5}$$

where w_{tr} is the tooth width at its root, m, and S is the number of armature slots (or teeth).

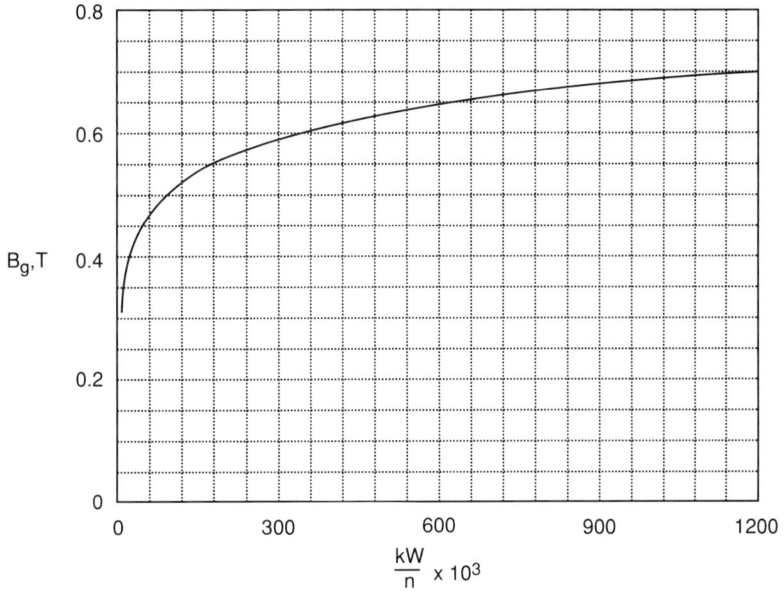

Figure 4.17 Air-gap flux densities for small d.c. motors.

Therefore, for a given number and size of slots, the tooth density is directly proportional to the maximum flux density in the air-gap. The value of air-gap flux density should be so chosen that the peak tooth density does not exceed 1.8 T. Increasing the tooth flux density above this value results in its magnetic saturation and a significant increase in the mmf required to send the flux through the teeth. This, in turn, places additional demand on the design of the excitation system. Furthermore, the iron losses in the teeth increases with the teeth density.

The air-gap flux density value used in designing a d.c. machine also depends on the speed of rotation. This is because as the machine rotates, the armature becomes magnetised alternately by north and south poles. The frequency of the flux reversal in the armature is $pn/2$; and this obviously affects the selection of the number of poles and is discussed later. In order to avoid excessive iron losses in the armature, low values of B_g should be used in high speed motors.

The value of B_g in small d.c. machines varies between 0.3 and 0.8 T; the lower values being for the smaller sizes. Initial values of B_g may be taken from the graph shown in Figure 14.17.

4.3.2 Electric loading

The value of specific electric loading (ac) of a d.c. machine is limited by the permissible temperature rise, commutation, efficiency and cost of construction; not always in this order. A higher value of ac results in a high temperature rise of the armature winding and, therefore, requires more expensive high class insulation materials. For example, the value of ac in a machine insulated with a class F material (155 °C) can be about 40% higher than if class A insulation (105 °C) is used. Also, the type of enclosure specified and the cooling method employed

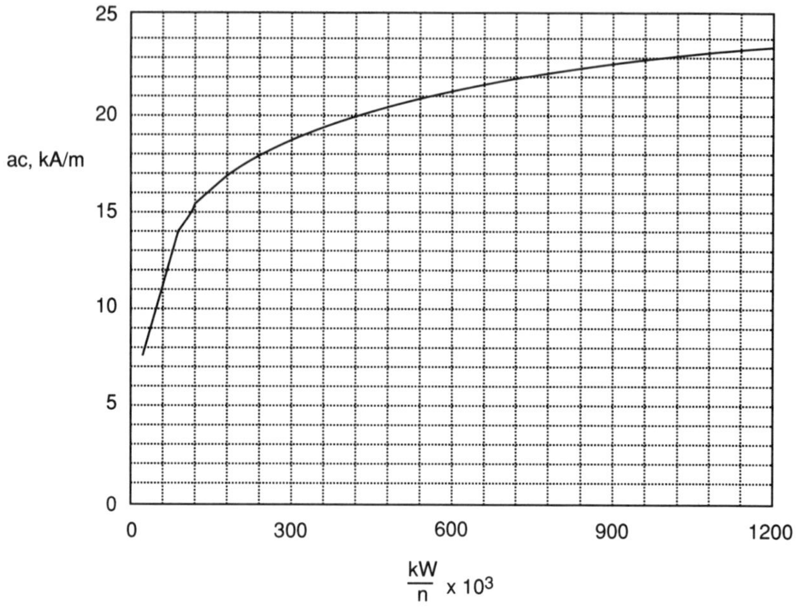

Figure 4.18 Specific electric loadings for small d.c. motors.

determine the temperature rise of a machine and, therefore, the design value of its specific electric loading.

The specific electric loading value (ampere-conductors per metre of the armature circumference), for a given rating, may be increased by increasing the number of armature conductors or by reducing the diameter. Increasing the number of conductor results in a high inductance of the armature coils (the inductance of the coil is proportional to the square of its number of turns), and the coil's reactance voltage will be increased. If a smaller diameter is used, the length of the core will have to be increased (to give required output) and narrow, deep slots will be used (to avoid tooth saturation). Both these changes result in an increase in the armature reactance and in the reactance voltage in the coil undergoing commutation (see *Electrical Machines and Their Applications* by J Hindmarsh, Pergamon Press), and this worsens the commutation conditions. Therefore, commutation is the main limiting factor in choosing the specific electric loading for machines without commutating poles. The limiting factors in choosing *ac* for machines provided with commutating poles are temperature rise, efficiency and cost. In the case of small motors, *ac* is usually from 8000 to 24,000 ampere-conductor/m. Initial design values of *ac* for commutating-pole machines may be taken from the graph shown in Figure 14.18 where average values are given for machines with class B insulation. Lower values may be used for a non-commutating pole machine.

It should be noted that both the values of B_g and *ac* may be considerably increased for intermittent duty motors.

4.3.3 Number of poles

Unlike alternating current machines, where the number of poles is determined by the supply frequency and the speed, a d.c. machine can be designed with various numbers of poles to give the same operating speed. However, there is a very narrow range of number of poles that yields satisfactory designs. The number of poles should be chosen, in the general case, so that the machine meets the design specifications with minimum weight of active material and minimum manufacturing cost.

The main limiting factor in selecting the number of poles is the frequency of flux reversal in the armature core (and currents in the armature conductors); which is equal to $pn/2$. For a given flux density, the losses in the armature core and teeth increase with the armature frequency. In general, the value of armature frequency varies between 25–50 Hz. In high speed applications, where even with two poles the frequency will be higher than 50 Hz, the flux density in the armature core and teeth must be kept low in order to avoid excessive core losses in the armature. This may be achieved either by increasing the iron area or by designing for a low value of B_g.

The pole pitch is given by:

$$\tau = \frac{\pi D}{p} \tag{4.6}$$

The pole pitch varies directly with the armature diameter and inversely with the number of poles. Increasing the number of poles for a given diameter results in a smaller pole pitch and a shorter length of end connections. The end connections do not take part in the energy conversion process. Therefore, increasing the number of poles results in a reduction of the armature copper losses and in the weight of copper used in the armature winding. Additionally, the cross-section area of the field pole (and the mean length of field coil turn) and pole ampere-turns vary directly with the pole pitch, or inversely with the number of poles for a given diameter. Increasing the number of poles for a given diameter results in a marginal reduction of the field copper weight and a significant reduction of field coil ampere-turns. Both these factors result in a reduction of the field copper losses and make it easier to ventilate the field coils. The designer has to weigh the advantages of increasing the number of poles against the increase in armature core losses.

The yoke's flux is inversely proportional to the number of poles. Therefore, some savings in the weight of the iron parts may be achieved be increasing the number of poles. Also, there are some other advantages arising from increasing the number of poles. These mainly result from the designer having more freedom in designing the armature winding and brush gear and commutator; but these factors are of little or no relevance to the subject of small d.c. motors.

The vast majority of small d.c. motors are designed with either 2 or 4 poles. It is usually desirable to select the number of poles that give a pole pitch less than 380 mm without exceeding 50 Hz in the armature.

4.3.4 Armature diameter and length

After selecting suitable values of specific electric and magnetic loadings, the output coefficient and the product D^2L are calculated from equations (4.2) and (4.1); respectively. The next step is to separate the product D^2L.

The main limitation in selecting the armature diameter is the rotor peripheral speed. Armature speeds above 30 m/s lead to expensive constructions and commutation problems. The peripheral speed of small d.c. motors should be between 8 and 25 m/s.

Using a long armature reduces the proportion of the inactive copper weight, used for the end connections, to the active conductor's weight. On the other hand, ventilating machines with large core length is difficult because the central section of the core tends to have a high temperature rise. Therefore, the armature length must be kept within certain limits.

The machine is usually proportioned by assuming a value of the length to pole pitch ratio as this approach takes into account some the factors which are discussed above regarding selecting the number of poles. For small d.c. motors, the ratio L/τ is usually between 0.6 and 0.9. In many cases it is necessary to make preliminary design calculations for different values of diameter and length and select the one which meets the design specifications at a reasonable manufacturing cost.

When designing a range of d.c. motors, the armature diameter should be so selected that as many ratings as possible can be realised by varying the stack length.

4.3.5 Air-gap length and pole face profile

The choice of the air-gap length is influenced by a number of factors. For example, increasing the air-gap serves to reduce distortion of the field form by armature reaction and to decrease the pulsation losses (caused by armature slots) in the pole faces; but this requires an increase in the field coil ampere-turns. Other advantages arising from increasing the radial length of the air-gap include quieter operation and better ventilation; but how far can a designer go in increasing the length of the air-gap and the field coil ampere-turns (or the length of pole magnet in the case of permanent magnet motors)? Satisfactory designs are usually obtained with air-gap values of between 1 and 1.5% of the pole pitch.

The air-gap length should be so selected that the pole does not become completely demagnetised at one end by the armature reaction field. The air-gap and armature teeth mmf per pole should not be less than the value of the armature reaction mmf under the pole tip. This determines the minimum air-gap radial length.

Good commutation is obtained when the air-gap flux density decreases gradually from the centre of the pole and reaching zero as the interpolar axis is reached. An air-gap flux distribution curve (field form) that drops off abruptly from maximum value to zero causes commutation difficulties and may give rise to magnetic noise.

In order to obtain the desired field form, the pole face is so shaped that the air-gap increases gradually with distance from the centre of the pole to the pole tips.

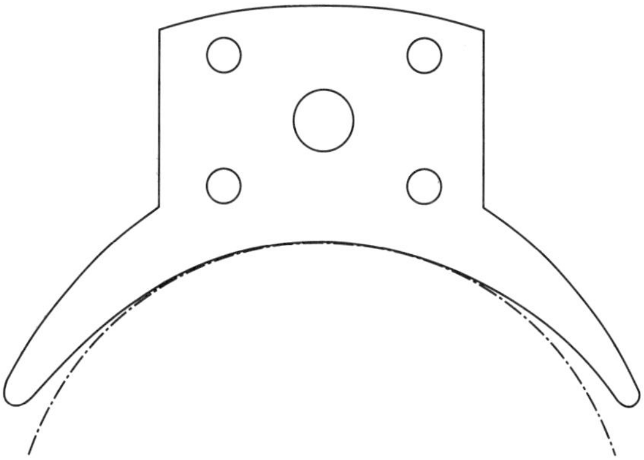

Figure 4.19 Pole face profile.

The air-gap length at the pole tips is generally about 1.5 times the air-gap at the centre of the pole; as illustrated in Figure 4.19.

4.3.6 Armature winding

The choice of armature winding for small d.c. motors is generally between simplex lap and simplex wave with the latter type preferred in most cases.

The number of conductors in a simplex lap winding is $p/2$ times that of a simplex wave winding. The cross-sectional area of a conductor used in a simplex wave winding is $p/2$ times the cross-section of the conductor required for the lap winding for the same machine. The smaller number of conductors with larger cross-section required for the wave winding results in better utilisation of slot area. This is because the proportion of the total slot space occupied by insulation falls as the number of conductors is reduced and as their cross-section is increased. Another advantage of the wave winding is that it does not require equaliser connections as unequal flux under the various poles has the same effect on the voltage induced in each armature path. The absence of equaliser connections and the better slot utilisation of the wave winding results in lower manufacturing cost.

The obvious advantage of the lap winding is that it permits reduction of the current per parallel path. Bearing in mind the threshold value of current per parallel path above which commutation difficulties arise is 250 A, a lap wound armature is highly unlikely to be employed in a small motor on this account. However, lap windings are used in some traction motors in order to reduce the motor's size and weight. This is because such a winding can be short-pitched which results in shorter overhang length at both the front and back.

The armature winding must be selected and designed with regard to the maximum voltage between adjacent commutator segments, to prevent flashover. Small motors rarely flashover when the maximum voltage between commutator segments is below 60 V. The designer should aim to limit the voltage between adjacent segments during all operating conditions to 30 V.

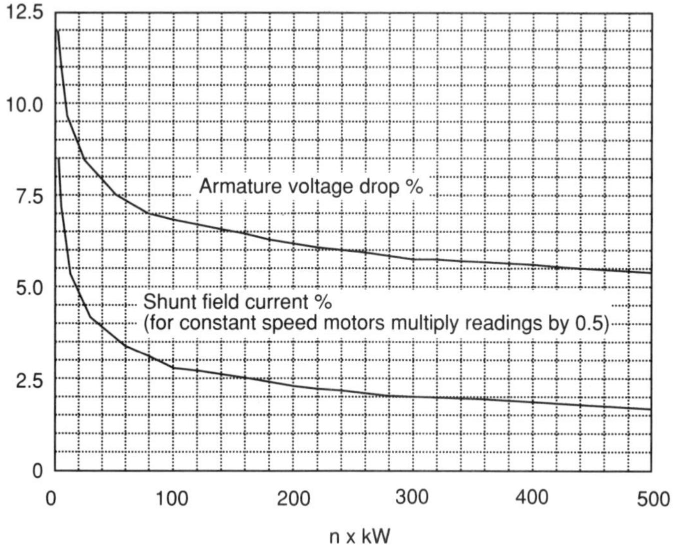

Figure 4.20 Approximate shunt-field current and armature circuit voltage drop as a percentage of rated current and terminal voltage, respectively.

The maximum voltage between adjacent commutator segments at no load and with load for machines without compensating windings is:

$$E_c = 2T_c N_c B_g L v_a \tag{4.7}$$

where T_c is the number of turns per coil, N_c is the number of coils connected between adjacent segments (which is equal to 1 for simplex wave and half the number of poles for simplex lap winding), B_g is the maximum flux density in the air-gap at no load, T, L is the armature length, m, and v_a is the armature peripheral speed, m/s.

For non-compensated machines, the field flux is distorted by the armature reaction field. The maximum flux density in the air-gap at full load could be as much as 25% higher than B_g.

The number of armature conductors (Z) is determined from the armature emf (E), speed (n), number of parallel paths (a), the pole flux (ϕ) and the number of poles (p) as follows:

$$Z = \frac{Ea}{\phi np} \tag{4.8}$$

To calculate the armature emf at the initial design stage, we need to estimate the armature voltage drop. This may be done with the aid of Figure 4.20 which gives the armature circuit voltage drop as a percentage of the terminal voltage.

The current per path in the armature winding for shunt motors is:

$$I_z = \frac{I_a}{a} = \frac{I - I_f}{a} \tag{4.9}$$

where: I_a is the armature current, I is the full load current of the motor and I_f is the shunt-field current.

The shunt-field current may be estimated from Figure 4.20 where I_f is given as a percentage of I.

Having estimated the armature conductor current (I_z), we assume a value for the current density and calculate the cross-section area of the conductor from:

$$a_z = \frac{I_z}{\delta_a} \tag{4.10}$$

where δ_a is the current density, A/mm^2, and a_z is the conductor cross-section area, mm^2.

The highest possible current density that gives a motor meeting the design specifications and satisfying the thermal design considerations should be selected. This not only leads to saving in armature copper but also enables shallow armature slots to be used, which aids commutation.

The usual current density value for small totally enclosed motors varies between 3 and 7 A/mm^2, the higher values are used for high speed fan-ventilated machines. The current density value should be considerably increased for intermittent duty motors.

It is always possible to meet the requirements of small d.c. motors with enamelled round copper conductors. The standard wire tables are given in Appendix 1.

The armature windings of almost all small d.c. motors are made on high speed winders and are transferred mechanically to pre-insulated slots. Sometimes, doubly-coated wires are used, with the top coating being abrasion resistant.

The length of the mean turn of the armature may be estimated from the following relationship:

$$L_{ma} = 2L + 2.3\tau + 5d_s \tag{4.11}$$

where L is the armature axial length, m, τ is the pole pitch, m, and d_s is the slot depth, m.

There are Z/a conductors in each parallel circuit and two coil sides per turn. The armature resistance (R_a) is $1/a$ times the resistance of each parallel circuit. The resistance of the armature winding is:

$$R_a = \frac{Z}{2a^2} \frac{\rho L_{ma}}{a_z \times 10^{-6}} \tag{4.12}$$

where a_z is the copper cross-section area in the conductor, mm^2, ρ is the resistivity of copper, Ωm and L_{ma} is in metres.

The resistance of the armature winding can be obtained using equation (4.12) at a given temperature corresponding to that at which the resistivity value is obtained (usually 25 °C). It must be corrected to obtain the value corresponding to the operating temperature.

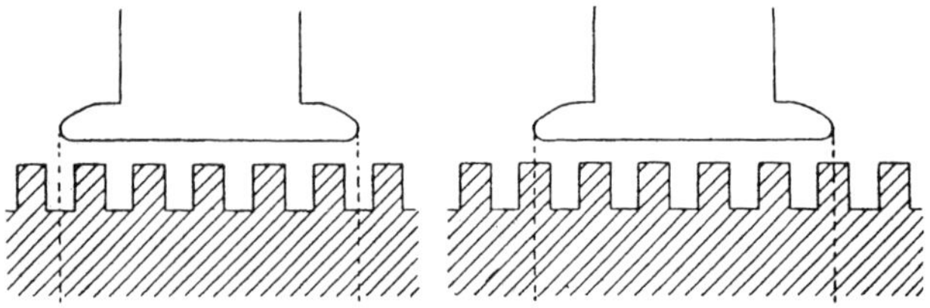

Figure 4.21 An illustration of the dependence of air-gap reluctance on armature position.

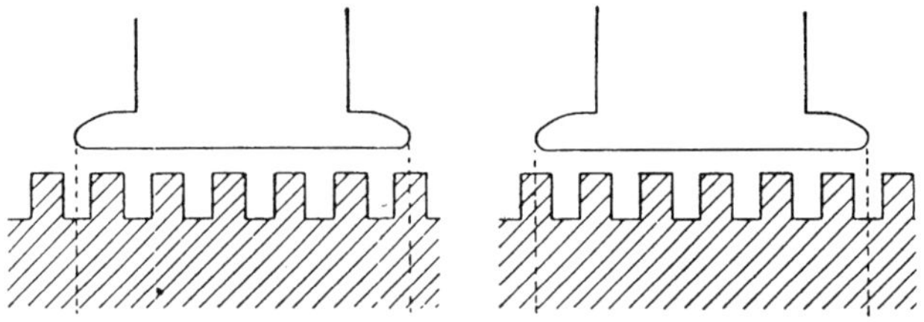

Figure 4.22

4.3.7 Armature slots

Having obtained approximate values of the number of armature conductors and their cross-section area, we now need to accommodate the conductors in the armature slots. As the armature windings of small d.c. motors always have two layers, the number of conductors per slot must be an even integer.

Choosing a small number of slots would result in saving in insulating material. This saving, however, will be at the expense of poorer commutation and excessive flux pulsation.

Flux pulsation is due to the variation of the reluctance of the flux path resulting from armature rotation. In Figure 4.21, the pole face covers 5 armature teeth and, in the armature position shown in Figure 4.21(a), the pole flux passes through 5 teeth. When the armature rotates half a tooth pitch the pole flux passes through 6 teeth; as illustrated in Figure 4.21(b). The reluctance of the air-gap changes with armature position when the number of slots per pole is a whole number. The air-gap flux pulsates when the armature rotates, and these pulsations produce iron losses in the pole face and give rise to magnetic noise.

When the number of slots under the pole face is made equal to 5 $\frac{1}{2}$, the total reluctance of the flux path and the flux under the pole remain approximately constant for all armature positions; as is illustrated in Figure 4.22. However, the reluctance and the flux under the pole tips vary with armature position. Armature rotation gives rise to flux oscillation under the pole tips and to ripple voltage in the conductors moving under the pole.

To avoid pulsations and oscillations of the air-gap flux, the number of slots per pole should be chosen as a whole number plus $\frac{1}{2}$. Such an arrangement makes the reluctance of the flux path per pole pair practically constant for all armature positions, as is shown in Figure 4.23.

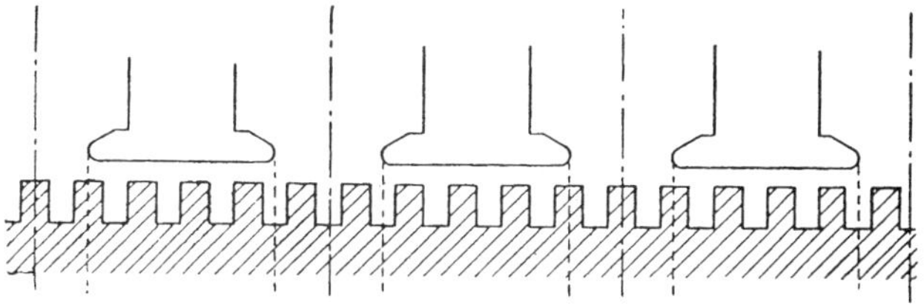

Figure 4.23

Pulsations and oscillations of the flux under the interpoles must be avoided, as they cause sparking and damage to the commutator. A large air-gap under the interpole and a small tooth pitch help to reduce the effect of slotting upon the flux under the interpole. In general, the number of slots between the tips of adjacent poles should be at least 3, or

$$(1 - \psi)\frac{S}{p} \geq 3 \tag{4.13}$$

where ψ is the pole arc to pole pitch ratio, S is the number of armature slots and p is the number of poles.

Assuming ψ between 0.66 to 0.7, the number of slots should be equal to or greater than between 8.8 and 10. For small motors with small armature diameter it may not be possible to use such a large number of slots per pole. In such a case, partly closed armature slots will be used. The usual value for slot pitch in small d.c. motors is between 15 to 25 mm, but for some very small motors it may become necessary to use a value less than 15 mm.

Another important factor which needs to be examined when selecting the number of slots is its suitability for the winding as regards the total number of coils and coil sides per slot.

The number of slots should be so chosen that the number of conductors per slot is an even integer. Also, the number of conductors per slot should be divisible by the number of coil sides per slot.

In case of a lap winding, the number of slots should be a multiple of the number of pole pairs. This makes the winding symmetric with respect to the poles and permits use of equalising connections. For a wave winding, the number of slots should not be a multiple of pole pairs otherwise dummy coils will be required.

Having decided on the number of armature slots the slot (or tooth) pitch is calculated from:

$$\tau_s = \frac{\pi D}{S} \tag{4.14}$$

The slot pitch is then proportioned so that the resulting slots have enough space to accommodate the armature coils together with all insulating material, and the tooth is wide enough to carry the flux without becoming saturated. In most cases, tapered slots are used in small d.c. motors.

If deep slots are used, their permeance coefficients will be high and, consequently, the reactance voltage will be high. Therefore, the depth of slots used in motors not provided with interpoles is limited to about 4 times its maximum width; with due consideration being given to the tooth width. Motors with up to 150 mm armature diameter are rarely designed with slots of more than 22 mm in depth. Deeper slots, however, may be used for machines having interpoles (as the interpoles produce a commutating field to compensate for reactance voltages).

4.3.8 Commutator and brushes

The number of commutator segments of a d.c. machine is always equal to the number of armature coils. Therefore, once the armature winding is determined, the number of segments becomes known. The minimum number of commutator segments (or armature coils) is determined by the voltage between adjacent segments; this value should not exceed 15 V at no load.

The commutator diameter is usually between 0.6 to 0.8 of the armature diameter. Generally, for a given rating, a larger commutator diameter will be required as the motor's voltage rating is decreased. This is because the armature current is higher for low voltage motors and this results in higher mechanical stresses and heating. Other factors which need to be examined when deciding on a value for the commutator diameter are the commutator peripheral speed, the distance between the brush arms and the commutator segment pitch.

The commutator peripheral speed should be kept below 15 m/s as a higher speed worsens the commutation conditions.

The voltage gradient along the commutator periphery should be kept to below 3 V/mm otherwise the air skin at the commutator surface may become ionised. For example, the minimum distance between the brush arms along the commutator surface of a 220 V motor should be 74 mm.

The commutator segment pitch (β_c) and the brush thickness determine the width of the commutation zone and the number of coils undergoing commutation. Commutation difficulties arise if the commutator segment pitch is less than 4 mm; therefore:

$$\beta_c = \frac{\pi D_c}{C} \geq 4\,\text{mm} \tag{4.15}$$

where D_c is the commutator diameter, m and C is the number of armature coils (or commutator segments).

If the thickness of the insulating spacer is about 0.8 mm, the thickness of the commutator segment at the surface should not be less than about 3 mm. The brush thickness should be so selected that it covers two to three commutator segments. The brush area should be selected so that it carries not more than 70 A. The requirements of small d.c. motors are usually met by just one brush on the brush arm.

Table 4.1 Properties of brush materials

Material type	Contact drop (V)	Current density (A/mm^2)	Pressure (kN/m^2)	Coefficient of friction
Natural graphite	0.7–1.2	0.1	14	0.1–0.2
Hard carbon	0.7–1.8	0.065–0.085	14–20	0.15–0.25
Electro-graphite	0.7–0.8	0.08–0.1	18–31	0.1–0.2
Metal graphite	0.4–0.7	0.1–0.2	18–21	0.1–0.2

The length of the commutator is determined by the space required to accommodate the brushes and the surface area required to dissipate the heat produced by its losses. The commutator length of small motors is usually equal to the brush axial dimension plus between 10 to 25 mm. Longer commutators may be used if the surface area, resulting from calculating the commutator length in this way, is too small to permit effective cooling.

Losses at the commutator surface result from the voltage drop at the brush/commutator interface (contact drop) and from friction losses. Average values of brush contact drop, current density and coefficient of friction for different brush materials are given in Table 4.1.

It should be noted that although the brush contact drop is independent of load current, the contact loss depends on the commutator condition and on the quality of commutation obtained. Therefore, it is very difficult to predetermine accurately the brush contact losses.

The friction losses depend on the brush pressure, on the commutator peripheral speed and on the coefficient of friction between commutator and brush. The coefficient of friction is a function of speed; decreases as speed is increased. The friction losses may be calculated from:

$$W_{bf} = \mu A_B P_b v_c \tag{4.16}$$

where W_{bf} is the friction loss, W, μ is the dimensionless coefficient of brush friction, A_B is the total contact area of all brushes, m^2, P_b is the brush contact pressure on the commutator surface, N/m^2, and v_c is the commutator peripheral speed, m/s.

4.3.9 Pole and yoke dimensions

The cross-section area of the pole is limited by the maximum flux density in the pole body and the pole height is selected so that the field coils may be accommodated and cooled effectively.

Because of leakage, the flux in the pole body may be estimated to be between 10 and 20% higher than the flux per pole value required to produce the armature emf (the smaller the machine the higher the leakage flux). The flux density in the pole body may be assumed to be between 1.2 and 1.8 T; depending on whether solid or laminated poles are used and on the quality of the laminations. The pole axial length (l_p) may be assumed equal to the armature length (L), and the length of the active material of a laminated pole (l_{pi}) may be assumed to be 0.95 times the mechanical length. Therefore:

$$A_p = b_p l_{pi} = \frac{\phi_p}{B_p} \tag{4.17}$$

and,

$$b_p = \frac{A_p}{l_{pi}} \tag{4.18}$$

where: ϕ_p is the flux in the pole body, Wb, B_p is the flux density in the pole body, T, and b_p is the width of the pole body (pole breadth), m.

The value of pole ampere-turns at full load determines the pole height (h_p), and it becomes known when the open circuit characteristics (magnetisation curve) of the motor is known. To construct the magnetisation curve of the machine we need to know the mmf required to pass the flux in the pole and yoke. Therefore, the field mmf and the height of the pole are interrelated and neither is known at this stage.

The ampere-turns per pole required to pass the flux in the pole and yoke may be, initially, estimated as 2000 times the pole pitch.

As well as affecting the pole mmf, the pole height influences the design of the field coil. The field conductors must be so arranged that effective cooling of the field coil is obtained. For a given number of pole ampere-turns, the thickness of the field coil increases, and its surface area decreases, as the pole height is reduced. An increase in the coil thickness results in a larger difference between the temperatures of its inner and outer surfaces, as heat is dissipated from the outer surface which is exposed to ambient air. Obviously, reduction of the coil surface area, resulting from shorter pole height, increases the temperature rise in the field coil. On the other hand, if the pole height is increased above a certain limit, the size of the motor will be unnecessarily increased.

The pole height may be taken to between 1 and 1.5 times the width of the pole body for rectangular poles. If solid circular poles are used, the pole height may be taken equal to half the diameter of its cross-section.

The leakage coefficient for yokes is slightly higher than that for poles. However, at the initial design stage we may assume that the yoke carries half the pole flux; therefore:

$$\phi_y = \frac{\phi_p}{2} \tag{4.19}$$

The yoke depth (d_y) is then found by assuming a value for the yoke's flux density. The flux density of the yoke may be assumed to be 1.2 T for cast steel yokes and between 1.5 and 1.6 T for laminated yokes. Therefore:

$$d_y = \frac{\phi_p}{2B_y l_{iy}} \tag{4.20}$$

where: B_y is the yoke's flux density, T and l_{iy} is the yoke's nett axial length, m.

The nett axial length of the yoke is equal to the actual length in the case of cast steel yokes, and is equal to the nett iron length of the armature in the case of laminated yokes.

4.3.10 Field winding design

The field coils of small d.c. motors always use round enamelled copper conductors. In most cases, the coils are wound directly on the pole body. The insulation used between the pole body and the field coil is usually made from pressboard.

The pole mmf must be known before we can design the field coil. The field ampere-turns may be calculated by the method described in Appendix 2 when suitable values of the flux densities in the various parts of the motor are assumed. Alternatively, we may estimate the required field mmf in relation to the armature reaction mmf at this initial design stage. The field coil will then be designed to provide this mmf value and its temperature rise will be examined. The number of turns on the field coil will be adjusted at a later stage.

In order to reduce the effect of the armature field (armature reaction) on the excitation field, the field mmf should be comparatively higher than the mmf developed by the armature current at full load. The ratio of the field emf at full load (AT_{fl}) to the armature mmf at full load (AT_a) may be chosen between 1.1 and 1.25; depending on the duty requirements.

The field coil should be so designed that it occupies minimum space without its temperature rise exceeding a specified limit. The coil temperature rise is influenced by the conductor current density, the depth of the coil and its surface area which is in contact with the coolant. We need to consider these factors together when designing a field coil.

Generally, the depth (thickness) of the field coil (b_{fc}) should not exceed 40 to 50 mm. The field winding conductor current density (δ_{fc}) may be chosen between 1.2 and 2.5 A/mm^2 when no ventilating ducts are used; which is the case in most small d.c. motors. The permissible loss per unit area of the field winding surface (p_f) may be assumed between 500 and 750 W/m^2.

The length of the mean turn of the field winding is:

$$l_{fc} = 2(l_p + 2b_{fc} + b_p) \tag{4.21}$$

where l_p is the axial length of the pole, m, b_p is the width of the pole body, m, and b_{fc} is the depth of the field coil, m.

The power dissipated from the field coil surface per metre of the coil height, excluding the top and bottom surfaces, is:

$$p_f = 2l_{fc}p_f \qquad \text{W/m} \tag{4.22}$$

We now assume the copper space factor in the field winding (f_{cf}) to be between 0.4 and 0.65; the lower value for conductors with small cross-section area. The copper volume per unit length of the field coil height will be:

$$v_{cf} = b_{fc}l_{fc}f_{cf} \qquad \text{m}^3/\text{m} \tag{4.23}$$

The copper volume per unit field coil height may be expressed in terms of the number of turns per unit height, n_{fc}, the conductor cross-section area, a_{fc}, and the length of the mean turn of the field coil as follows:

$$v_{cf} = n_{fc}(a_{fc} \times 10^{-6})l_{fc} \qquad \mathrm{m^3/m} \tag{4.24}$$

where a_{fc} is the conductor cross-section area in mm^2 and the unit of n_{fc} is m^{-1}.

The field current I_f is:

$$I_f = \delta_{fc} a_{fc} \tag{4.25}$$

where δ_{fc} is the current density in the field coil, A/mm^2 and a_{fc} is the conductor cross-section area, mm^2.

The resistance of the n_{fc} turns is:

$$R_n = \frac{\rho n_{fc} l_{fc}}{a_{fc} \times 10^{-6}} \qquad \Omega/\mathrm{m} \tag{4.26}$$

where ρ is the resistivity of copper, Ωm.

The copper loss produced in the n_{fc} turns is $I_f^2 R_n$. This must equal to the power dissipated from the coil surface per metre of the coil height given in equation (4.22), therefore:

$$2l_{fc}p_f = \delta_{fc}^2 a_{fc}^2 \frac{\rho n_{fc} l_{fc}}{a_{fc} \times 10^{-6}} \tag{4.27}$$

Utilising equations (4.24) and (4.23), equation (4.27) is rewritten as:

$$2l_{fc}p_f = \delta_{fc}^2 \rho b_{fc} l_{fc} f_{cf} \times 10^{12} \tag{4.28}$$

The current density in the field conductor will be:

$$\delta_{fc} = \sqrt{\frac{2p_f \times 10^{-12}}{\rho b_{fc} f_{cf}}} \tag{4.29}$$

where the unit of p_f is W/m^2, the unit of ρ is Ωm, the unit of b_{fc} is m and the unit of δ_{fc} is A/mm^2.

The ampere-turns per metre height of the field coil is:

$$\frac{AT_{fl}}{h_f} = I_f \frac{b_{fc} f_{cf}}{a_{fc} \times 10^{-6}} = \delta_{fc} b_{fc} f_{cf} \times 10^6 \tag{4.30}$$

where AT_{fl} is the pole ampere-turns at full-load, a_{fc} is the cross-section area of the field conductor, mm^2 and h_f is the height of the field winding, m.

Equation (4.30) may be rearranged to give the height of the field coil as follows:

$$h_f = \frac{AT_{f1}}{\delta_{fc} b_{fc} f_{cf} \times 10^6} \tag{4.31}$$

Substituting the value of the conductor current density from equation (4.29) into equation (4.30) we get:

$$h_f = AT_{f1} \sqrt{\frac{\rho}{2p_f b_{fc} f_{cf}}} \tag{4.32}$$

The pole height, h_p, excluding the height of the pole shoe, h_s, should be taken slightly greater than the height of the field coil, h_f. This is to allow for the space taken by insulation and the curvature of the yoke. This increase may be taken between 5% and 10% of the pole pitch.

Shunt-field windings in small d.c. motors employ coils of several hundred turns of small section enamelled copper wire. In shunt motors the entire winding space is taken up by the shunt-field winding. In compound motors about 20% of the winding space is taken by the series field winding.

In constant speed, or armature controlled, motors the rated voltage is absorbed by the shunt-field winding, and the voltage across each field coil is the terminal voltage divided by the number of poles. In the case of field controlled motors, an allowance is made for the voltage absorbed across the field rheostat; depending on the range of speed control.

The resistance of the field winding is:

$$R_f = \frac{V_f}{I_f} = pN_f \frac{\rho l_{fc}}{a_{fc} \times 10^{-6}} \tag{4.33}$$

where V_f is the voltage applied across the field winding, N_f is the number of turns in each field coil, a_{fc} is the conductor cross-sectional area in mm^2 and ρ is the resistivity of copper in Ωm.

The cross-section area of the shunt-field conductor is:

$$a_{fc} = \frac{pI_f N_f \rho l_{fc}}{V_f \times 10^{-6}} = \frac{pAT_{f1} \rho l_{fc}}{V_f \times 10^{-6}} \text{ mm}^2 \tag{4.34}$$

Equation (4.34) may be rearranged to give the number of ampere-turns per pole as follows:

$$I_f N_f = \frac{V_f a_{fc} \times 10^{-6}}{p \rho l_{fc}} \tag{4.35}$$

Equation (4.35) shows that the number of ampere-turns per pole on the shunt-field winding is independent of the number of turns per pole, but it varies directly with the cross-section area of the conductor.

Since the number of turns per pole on the field winding determines the heating, we need to calculate at this point the loss per unit area of the field winding (p_f) and compare it with the permissible values of between 500 and 750 W/m^2. The area of the cooling surface is taken equal to the perimeter of the coil section times the length of the mean turn. The total area of cooling surface for p poles is approximately:

$$A_{sf} = 2(l_{fc} + b_{fc})h_f p \tag{4.36}$$

and the power generated per unit area is:

$$p_f = \frac{I_f^2 R_f}{A_{sf}} \leq 500 \text{ to } 750\,\text{W/m}^2 \tag{4.37}$$

Series field

The mmf to be provided by the series field at full load is usually between 15 and 25% of the armature mmf. The number of series field turns is readily determined by dividing its ampere-turns by the value of armature current. The conductor cross-section area is determined by assuming current density value between 2 and $3\,\text{A/mm}^2$.

4.3.11 Interpoles

The purpose of the interpole is to provide flux so that the rotational emf it produces in the coil undergoing commutation cancels out the reactance emf in the coil. Almost all d.c. motors of more than 0.75 kW are provided with interpoles or commutating poles.

The commutating pole winding is connected in series with the armature, and it carries the total armature current. Round copper conductors are used for the commutating pole winding of small d.c. motors.

The interpoles may be made of cast steel or punched from steel sheet with no special pole shoe. However, the modern trend is use completely laminated cores, and the length of the interpole is usually equal to that of the main pole.

The width of the interpole must be equal to the width of the commutating zone; an allowance of between 1.5 to 2 times the air-gap under the interpole may be made for the fringing flux at the tips of the interpole. The width of the commutation zone for wave wound motors is:

$$w_c = \frac{D}{D_c}\left[\left(\frac{u}{2} - \frac{a}{p}\right)\beta_c + (w_b - w_m)\right] \tag{4.38}$$

where D and D_c are the armature and commutator diameter, respectively, u is the number of coil sides per slot, β_c is the commutator pitch, m, and w_b and w_m are the width of the brush along the commutator circumference and width of the insulating spacer, respectively, in metres.

The commutator pole width must also be so chosen such that the leakage flux will not be excessive. To avoid excessive leakage, the width of the commutating pole should not exceed half the space between adjacent pole tips; therefore:

$$w_{ip} \leq \frac{1 - \psi}{2} \tau \tag{4.39}$$

The interpole width must also be chosen with regard to armature tooth pitch. With a narrow interpole, the pulsations of its flux caused by armature slots will be large. The width of the interpole should be larger than 1.5 times the tooth pitch, and if possible, should be a multiple of the tooth pitch. Therefore:

$$w_{ip} \leq 1.5\, \tau_s \tag{4.40}$$

The air-gap under the commutating pole is generally between one to two times the air-gap under the main pole.

The ampere-turns on the interpole are required to provide mmf to establish commutating flux in the air-gap and mmf to overcome the armature reaction ampere-turns at its strongest point. The ampere-turns required to furnish the commutating flux can be calculated if the required flux density under the interpole, B_{gi}, is known. B_{gi} can be found if the reactance voltage developed in the coil undergoing commutation is known. The interpole ampere-turns are usually between 1.2 and 1.3 times the armature ampere-turns; therefore:

$$AT_{ip} = (1.2 \text{ to } 1.3) \times \frac{I_z Z}{2p} \tag{4.41}$$

The number of interpole turns is:

$$N_{ip} = \frac{AT_{ip}}{I_a} \tag{4.42}$$

The current density in the commutating winding, δ_{ip}, may be assumed between 1.2 and 2.5 A/mm^2. The cross-section area of the conductor is then obtained as follows:

$$a_{ip} = \frac{I_a}{\delta_{ip}} \text{ mm}^2 \tag{4.43}$$

The length of the mean turn of the commutating field coil, l_{ic}, can be found in terms of the width of the interpole, its axial length and the coil thickness following the same procedure used for the shunt-field coil. The total resistance of the commutating field winding is:

$$R_{ip} = \frac{\rho l_{ic} N_{ip}}{a_{ip} \times 10^{-6}} p \tag{4.44}$$

4.3.12 Losses and efficiency

The losses in a d.c. motor are:

(1) Copper losses in the armature and field windings.

(2) Brush contact I^2R losses.

(3) Core or iron losses in armature teeth and core.

(4) Field and armature rheostat losses (when present).

(5) Mechanical losses due to friction (bearings and brushes) and windage.

In addition to the losses given above, there are additional load losses which may be of importance. These losses include eddy current losses in armature conductors, losses in the brushes caused by commutation and iron losses introduced by load. Because there are no satisfactory methods available for determining these losses, the IEC Standards recommend that they be taken equal to 1% and 0.5% of the rated input, for uncompensated and compensated motors; respectively.

In accordance with the International Standards, the copper losses for all windings are to be calculated for a temperature of 75 °C for Classes A, E and B, and for a temperature of 110 °C for Classes F and H; for all loads.

Armature copper loss

The resistance of the armature winding is given in equation (4.12). The armature copper loss is:

$$W_a = I_a^2 R_a \tag{4.45}$$

For a motor, the armature current is the terminal current minus the shunt-field current. The armature copper loss varies with the square of the load. It should be noted that the armature resistance used in equation (4.45) is the resistance corrected to the reference temperature (75 or 110 °C, depending on the Class of the machine).

Commutating field copper losses

The commutating field winding (and the compensating winding), when used, carry the armature current. The resistance of the commutating field winding is given in equation (4.44). The copper loss in the commutating field winding varies with the square of the load, and is given by:

$$W_i = I_a^2 R_{ip} \tag{4.46}$$

Shunt field copper loss

When a rheostat is connected in series with the field winding, the rheostat loss must be included with the shunt-field copper loss when calculating the efficiency. The shunt-field resistance is given by equation (4.33), and the shunt-field copper loss is:

$$W_f = I_f^2 R_f \tag{4.47}$$

Brush contact loss

The brush contact I^2R loss depends on the condition of the commutator and on the quality of the commutation obtained. Therefore, this loss is very difficult to predetermine accurately. IEC Standards recommend that 2 V be assumed for the total drop at the brush contact (i.e. for both positive and negative brushes), for carbon and graphite brushes. The recommended value of the total voltage drop is 0.6 V for metal graphite brushes. Therefore, depending on the type of brushes used, the brush contact I^2R loss is:

$$W_b = (2 \text{ or } 0.6)I_a \tag{4.48}$$

Core loss

The iron losses in electric sheet steel consist of the hysteresis and the eddy current losses. In general, the flux density is not the same for the armature teeth and core, and their core losses must be calculated separately. Specific core loss values (loss per kg) due to the fundamental frequency flux and corresponding to the working flux density can be obtained from the curves given in Appendix 1. These values must be corrected to the frequency of flux reversal in the armature (or $pn/2$).

The flux density in the armature core may be assumed constant over its cross-sectional area. In general, the flux density is not the same for all sections of the armature teeth because of the tooth taper. The iron loss in the armature teeth is calculated for the density at a section $\frac{1}{3}$ slot depth from the minimum tooth width. It should be noted that the average tooth width is always used when calculating the weight of the armature teeth.

The specific loss values when multiplied by the frequency correction factor and the weight of iron then give the iron loss in Watts.

The weight of the armature teeth in kilograms is:

$$G_t = S w_t l_{ia} d_s \gamma \tag{4.49}$$

where S is the number of slots, w_t is the average tooth width, m, l_{ia} is the armature iron length, m, d_s is the slot depth, m, and γ is the specific weight of the lamination material, kg/m^3. The weight of the armature core in kilograms is:

$$G_c = \frac{\pi}{4}[(D - 2d_s)^2 - D_i^2] l_{ia} \gamma \tag{4.50}$$

where D and D_i are the outer and inner diameters of the lamination, respectively, in metres.

In addition to the hysteresis and eddy current losses in the laminated armature produced by the fundamental frequency flux, there are losses in the pole face, due to flux pulsations, losses due to imperfect insulation between the laminations (caused by burrs or slot filing) and losses due to punching and bending strains in the laminations. The total iron losses may be estimated by multiplying the sum of armature teeth and core losses (due to the fundamental frequency flux) by a factor which is generally between 1.5 and 3. This factor should be determined from test data for similar motors. When such data is not available, 2.5 may be used.

Brush friction loss

The brush friction loss depends on the brush pressure, the coefficient of friction between commutator and brush and commutator peripheral speed. An approximate value for the brush friction loss may be obtained from equation (4.16); using contact pressure and friction coefficient values from Table 4.1. However, tests on new machines show wide variations of brush friction loss, because the commutator and brushes do not have the smooth surfaces that come after continued operation. Test data obtained from a large number of machines show that the brush friction loss is approximately:

$$W_{bf} = 2400\ A_B v_c \quad \text{for carbon and graphite brushes}$$

$$W_{bf} = 1500\ A_B v_c \quad \text{for carbon graphite brushes}$$

where A_B is the total brush contact area, m^2 and v_c is the commutator peripheral speed, m/s.

Bearing friction and windage losses

Like brush friction, the bearing friction depends upon the pressure on the bearing, the peripheral speed of the shaft at the bearing and the coefficient of friction between bearing and shaft.

The windage loss depends on the rotor peripheral speed, on the rotor dimensions and on the construction of the machine.

The bearing friction and windage losses are generally not separated and are determined from tests on similar machines. When such data is not available, they can be estimated as a percentage of the rated output. In the case of a small d.c. machine, the bearing friction and windage losses may be taken equal to $v_a/50\%$ of the rated output, where v_a is the armature peripheral speed in metres per second.

4.4 SAMPLE DESIGN

In the following, the application of the design procedure described above to a typical design problem is demonstrated. We will work within typical constraints which would face a designer, such as existing laminations and availability of standard wires and brushes. At every stage the factors affecting the design decision are discussed.

Let us consider the problem of designing a constant speed d.c. motor rated at 1.5 kW, 2700 rev/min, 110 V.

4.4.1 Main dimensions

$$n = \frac{N(\text{rev/min})}{60} = \frac{2700}{60} = 45\,\text{rev/s}$$

$$\frac{\text{kW output}}{n} \times 10^3 = \frac{1.5}{45} \times 10^3 = \frac{1.5}{45} \times 10^3 = 33.3$$

From Figures 4.17 and 4.18 we can find B_g and ac; respectively:

$$B_g = 0.42\,\text{T}, \qquad ac = 9\,\text{kA/m}$$

Assume a pole arc to pole pitch ratio of 0.7; the average gap density (B_{av}) will be 0.294 T. The output coefficient is then calculated from equation (4.2):

$$C_o = \pi^2 \times 0.294 \times 9000 \times 10^{-3} = 26.11$$

The D^2L Product is then calculated:

$$D^2L = \frac{P_a}{C_o n} = \frac{1.5}{26.11 \times 45} = 1.276 \times 10^{-3}\,\text{m}^3$$

Assuming a core length to diameter ratio of 1.3 we get:

$$D = 100\,\text{mm}, \qquad L = 128\,\text{mm}$$

Now we need to examine the core length to pole pitch ratio and the armature peripheral speed.

The rotor speed is 45 rev/s, if we choose a number of poles higher than 2 the armature frequency will be high and so will be its core losses. Choosing a 2-pole machine, the pole pitch is:

$$\tau = \frac{\pi D}{p} = \frac{\pi \times 100}{2} = 157\,\text{mm}$$

Then:

$$\frac{L}{\tau} = \frac{128}{157} = 0.82$$

which is within the range (0.5 to 0.9) given in Section 4.3.4 above.

The armature peripheral speed is:

$$v_a = \pi D n = \pi \times 0.1 \times 45 = 14.12\,\text{m/s}$$

which is within the specified range of 8 to 25 m/s.

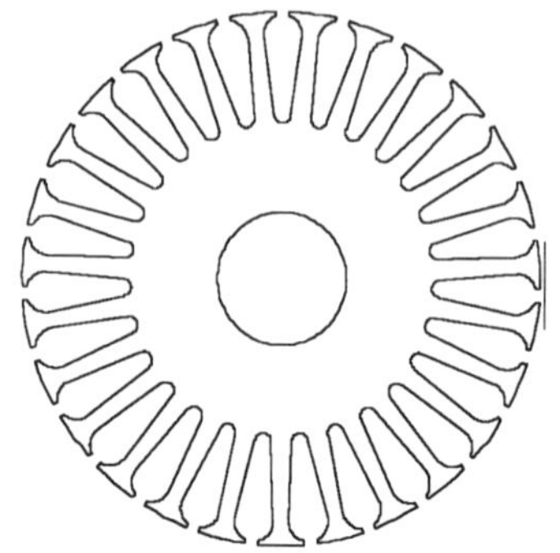

Figure 4.24 A 27-slot armature lamination.

4.4.2 Armature design

The average flux per pole is:

$$\phi = B_{av}\tau L = 0.294 \times 0.157 \times 0.128 = 5.9\,\text{mWb}$$

For motors, the no load voltage (open circuit induced emf), is equal to the terminal voltage. We need to estimate the no load speed. Let the no load speed be 2800 rev/min (or 46.67 rev/s). The total number of armature conductors is then calculated from equation (4.8):

$$Z = \frac{Ea}{\phi n_o p} = \frac{110 \times 2}{5.9 \times 10^{-3} \times 46.67 \times 2} = 399$$

where n_o is the no load speed. Although we have not yet decided on the type of armature winding, as the number of poles is 2, the number of parallel paths will be 2 whether we choose simple wave or lap. The induced voltage per conductor is:

$$e_c = \phi n_o = 5.9 \times 10^{-3} \times 46.67 = 0.275\,\text{V}$$

Since this is quite low, the simple wave winding can be used to advantage.

The number of slots between the pole tips should be at least 3. Therefore, this condition can be met with a number of slots equal to or greater than 20. An odd number of slots will yield slots per pole equal to a whole number plus $\frac{1}{2}$; which reduces flux pulsations in the air-gap. These conditions are satisfied with an odd number of slots equal to or greater than 21.

We should utilise existing laminations whenever possible. In this case let us assume that in an existing range of 100 mm diameter laminations there are 16, 27 or 36 slots. Although we only need 21 slots, it is economical to use existing laminations of higher numbers of slots. This will obviously increase the cost of slot insulations, but it eliminates the tooling cost of new laminations. Even if we did not have laminations with exactly the desired diameter we should utilise the nearest size of laminations and adjusted the stack length accordingly.

The 27-slot lamination is shown in Figure 4.24. Its slot area is 101 mm^2, core depth is 16.6 mm, tooth width is 3.5 mm and slot depth is 20.9 mm (including slot openings). We can only use this lamination if the flux density in the teeth is below the saturation value, if the number of conductors per slot can be made an even integer without having to change the specific electric loading significantly and if the slot area is large enough so that the number of conductors per slot can be accommodated together with conductor and slot insulation.

The tooth flux, ignoring slot leakage, is:

$$\phi_t = \frac{p\phi}{\psi S}$$

where S is the number of armature slots (27 in this case).

The flux density in the teeth is:

$$B_t = \frac{\phi_t}{w_t L_i}$$

where w_t is the narrowest tooth width and L_i is the armature iron length.

The length of armature iron may be taken equal to 0.95 of the armature length. Since the slots are tapered and the teeth have parallel sides, the tooth width is constant at 3.5 mm. Therefore:

$$B_t = \frac{2 \times 5.9 \times 10^{-3}}{0.7 \times 27 \times 3.5 \times 10^{-3} \times 0.95 \times 0.128} = 1.47\,\mathrm{T}$$

This is, in fact, somewhat low and one might consider increasing the specific magnetic loading value. First, let us examine the value of armature core flux density. Half the pole flux passes circumferentially in the armature core. The core depth of the lamination shown in Figure 4.24 is 16.6 mm and the resulting core flux density is about 1.46 T, which is marginally less than the maximum recommended value of 1.5 T.

It may be noted that if we were to design another motor with the same rated power but with a full load speed of, say, 1300 rev/min, we would have been able to design for 4 poles without exceeding an armature frequency of 50 Hz, and the resulting reduction of core flux would enable higher specific magnetic loading to be specified. This is one of the reasons why the B_g values in Figure 4.17 are plotted against the power to speed ratio. It is true that the specific loadings and the output coefficient of the 1300 rev/min motor will be higher than the corresponding values for the 2700 rev/min one. The D^2L product of the slower machine will, however, be higher than that of the faster one. This is because the product D^2L is inversely proportional to the product of speed and output coefficient; the increase in the output coefficient is always less than the reduction in speed. The diameter and length of the 1300 rev/min motor would be about 150 and 100 mm respectively.

The slot area of the laminations shown in Figure 4.24 is 101 mm^2. The number of conductors per slot is:

$$\frac{Z}{S} = \frac{399}{27} = 14.78$$

The number of coil sides per slot must be 2 or 6 in order to avoid using dummy coils in the wave wound armature, and the required number of conductors per slot must be a multiple of 2 or 6, respectively. If we take 12 conductors per slot we will be able to design the winding with either 2 or 6 coil sides per slot; but this will involve a significant reduction in the specific electric loading (which will result in a larger machine and/or higher specific magnetic loading). Therefore, in this case, we assume 2 coil sides per slot and we can use either 14 or 16 conductors per slot (7 or 8 turns per coil, respectively). Before we can finally decide to use the laminations shown in Figure 4.24 we must be satisfied that the 16 or 14 conductors can be accommodated in the slot; together with conductor and slot insulation.

The motor terminal current is:

$$I = \frac{P_o}{V\eta}$$

where P_o is the rated power, W, V is the terminal voltage, V and η is the efficiency, p.u.

Assuming an efficiency of 85% (0.85 p.u.), the motor terminal current will be 16.0 A.

From Figure 4.20, for $n \times \mathrm{kW}$ of 67.5, the shunt-field current should be approximately 1.34% of the rated motor current, or 0.2 A.

$$I_a = 16.0 - 0.2 = 15.8, \qquad I_z = \frac{I_a}{2} = 7.9\ \mathrm{A}$$

Assuming the conductor current density (δ_a) to be 3.5 A/mm^2, the conductor cross-section area is:

$$a_z = \frac{I_z}{\delta_a} = \frac{7.9}{3.5} = 2.26\ \mathrm{mm}^2$$

The conductor diameter is:

$$d_a = \sqrt{\frac{4 \times 2.26}{\pi}} = 1.69\ \mathrm{mm}$$

The nearest diameter in the standardised copper table is 1.7 mm (the copper area is 2.27 mm^2) and the diameter of insulated conductor is 1.83 mm. The cross-section area of the insulated conductor is about 2.6 mm^2. The total copper area of 16 is 36.3 mm^2. The slot area of the lamination shown in Figure 4.24 is 101 mm^2. This gives a slot space factor of 0.36. Therefore 16 (or, obviously, 14) conductors may be comfortably accommodated in the slot. In fact, the slot utilisation will be quite poor if we use these laminations. What we really need is laminations with smaller slots and deeper core. With such laminations, the ratio of copper to slot area will improve, providing better slot utilisation, and saturation of the core will be avoided.

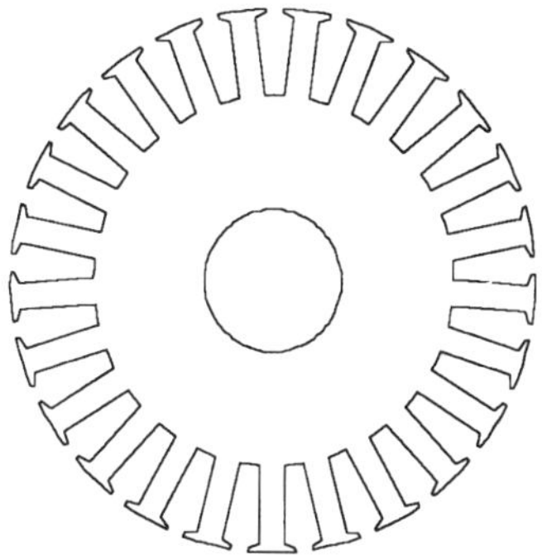

Figure 4.25 A 25-slot armature lamination.

Let us consider the lamination shown in Figure 4.25. Its outer diameter is 95 mm, slot area is 75.5 mm^2, core depth is 20 mm, tooth width is 4.5 mm and slot depth is 15 mm. The slot space factor will be 0.48 or 0.42 for 16 and 14 conductors per slot, respectively. The shorter slot depth (15 mm compared with 20.9 mm for the 27-slot lamination) will result in less ampere-turns being required to pass the flux in the teeth. The 25-slot lamination is therefore more suitable for this design.

We use the lamination of Figure 4.25, and with 14 conductors per slot, the specific electric loading is 9264 ampere-conductors per metre. The total number of armature conductors is 350, and the average flux per pole required with no load voltage equal to the terminal voltage of 110 V is 6.73 mWb. We choose the armature length so that the resulting flux density in the core is 1.5 T. This gives an armature length of 118 mm.

The maximum flux density in the air-gap, B_g, is now about 0.55 T, but the flux density in the teeth and core, at 1.5 T and 1.53 T respectively, are quite acceptable.

The ampere-turns per metre for both the tooth and armature core is found from the magnetisation curve of the lamination material given in Appendix 1. Therefore:

$$at_t = 700\ \text{A/m}, \qquad at_c = 800\ \text{A/m}$$

The slot depth, including slot opening, is 15 mm. The length of the flux path in the core, l_c, is calculated by equation (A2.15), Appendix 2, to be 35.4 mm. Therefore, the teeth and core ampere-turns per pole are:

$$AT_t = 700 \times 0.015 = 11, \qquad AT_c = 800 \times 35.4 = 28$$

There are 14 conductors per slot. The number of coil sides per slot is 2, and there are 7 turns per coil. This gives a number of armature coils and commutator segments equal to the number of slots, i.e. 25. we now examine the commutator pitch.

Assume the diameter of the commutator is 60% of the armature diameter, therefore:

$$\beta_c = \frac{\pi D_c}{C} = \frac{\pi \times 0.6 \times 95}{25} = 7.16\ \text{mm}$$

which is quite acceptable.

The commutator peripheral speed at no load is about 8 m/s, which is within the limits given in Section 4.3.8. The voltage between adjacent commutator segments, as calculated by equation (4.7), is about 12.2 V. Again this is satisfactory.

The length of the mean turn of the armature winding is obtained from equation (4.11) as follows:

$$L_{ma} = 2 \times 0.118 + 2.3 \times 0.149 + 5 \times 0.015 = 0.654\,\text{m}$$

and the armature resistance is obtained from equation (4.12):

$$R_a = \frac{350}{2 \times 2^2} \frac{1.8 \times 10^{-8} \times 0.654}{2.27 \times 10^{-6}} = 0.227\,\Omega$$

where the resistivity of copper is taken at 25 °C.

The armature resistance voltage drop is 3.6 V. This is about 3.3% of the terminal voltage. From Figure 4.20, for $n \times$ kW of 67.5, the armature circuit voltage drop is about 7.2%, or 7.9 V. After allowing for voltage drop in the commutating pole winding and for the brush contact drop, the armature circuit voltage drop will be about the value obtained from Figure 4.20. This shows that the armature design is satisfactory.

4.4.3 Air-gap and pole ampere-turns

In order that the motor's trailing pole tip does not become completely demagnetised, the ampere-turns per pole for the air-gap and armature teeth should be not less than the value of the armature reaction mmf at the pole tip, or $\psi\ AT_a$. Therefore, we take the ampere-turns required to pass the flux in the air-gap and armature teeth to be approximately 0.75 times the armature ampere-turns per pole.

The armature ampere-turns per pole is:

$$AT_a = \frac{I_z Z}{2p} = \frac{7.9 \times 350}{2 \times 2} = 691$$

and;

$$AT_g = 0.75\,AT_a - AT_t = 519 - 11 = 508$$

But:

$$AT_g = \frac{B_g}{\mu_0} l_g K_g$$

where K_g is the air-gap correction factor and is equal to 1.15 in this case (Figure A2.2, Appendix 2).

The air-gap length obtained in this way is about 1 mm. The ampere-turns per pole required to send the flux through the pole and yoke at full load, $(AT_p + AT_y)$, may be taken equal to 2000 times the pole pitch; where the pole pitch is in metres. Therefore:

$$AT_p + AT_y = 2000 \times 0.149 = 298$$

If we choose l_g equals to 1 mm the ampere-turns per pole at full load will be about 845; and this provisional value is about 22% greater than the armature reaction ampere-turns.

4.4.4 Pole and field coil design

To determine the pole dimensions, let us assume the flux density in the pole body to be 1.8 T for the laminated pole. The pole nett iron length is 0.95 times the armature length; therefore:

$$B_p = 1.8\,\text{T}, \qquad l_{pi} = 0.95 \times 0.118 = 0.112\,\text{m}$$

Assuming a pole leakage factor of 1.1, the flux in the pole body is:

$$\phi_p = 6.73 \times 1.1 = 7.4\,\text{mWb}$$

The pole cross-section area is:

$$A_p = \frac{\phi_p}{B_p} = \frac{7.4 \times 10^{-3}}{1.8} = 4.11 \times 10^{-3}\,\text{m}^2$$

The width of the pole body is:

$$b_p = \frac{A_p}{l_{pi}} = \frac{4.11 \times 10^{-3}}{0.112} = 0.036\,\text{m}$$

The pole height may be assumed between 1 and 1.5 times the pole width. Let the pole height, h_p, be 36 mm including the pole shoe. The ampere-turns per metre of the pole height is found from the magnetisation characteristics given in Appendix 1 to be 8000. The ampere-turns required to pass the flux in the pole body is:

$$AT_p = 8000 \times 0.036 = 288$$

Assuming the flux density in the yoke to be 1.5 T, we calculate the depth of the yoke from equation (4.20). The depth of the yoke is about 22 mm. The length of the flux path, l_y, is found with the aid of equation (A2.22), Appendix 2, to be 0.15 m. The ampere-turns per metre of the flux path in the yoke is 700. The ampere-turns required to pass the flux in the yoke is:

$$AT_y = at_y \times l_y = 700 \times 0.15 = 105$$

The ampere-turns per pole required to pass the flux in the pole body and yoke is therefore 393 (not 298 as estimated above), and the total ampere-turns per pole at full load is now 940.

The pole design will be satisfactory if the field coil can be accommodated. The height of the field coil is slightly less than the pole height. Let the average height of the field coil be 28 mm. Assuming the coil depth, b_{fc}, to be 40 mm, and the field coil space factor to be 0.55, equation (4.30) is used to calculate the current density in the field conductors. The resulting current density is about $1.52\,A/mm^2$. This value is acceptable and the pole dimensions are, so far, satisfactory.

The length of the mean turn of the field winding is calculated using equation (4.21):

$$l_{fc} = 2(0.118 + 2 \times 0.04 + 0.036) = 0.468\,\text{m}$$

The section area of the conductor is calculated using equation (4.34):

$$a_{fc} = \frac{2 \times 940 \times 1.8 \times 10^{-8} \times 0.468}{110} = 1.44 \times 10^{-7}\,\text{m}^2 = 0.144\,\text{mm}^2$$

The nearest standard wire has a diameter of 0.45 mm and its cross-section area is $0.159\,mm^2$. The diameter of the doubly insulated wire is 0.516. A more accurate value for the field coil space factor can now be obtained as follows:

$$f_{cf} = 0.75 \times \left(\frac{0.45}{0.516}\right)^2 = 0.57$$

which is very close to the assumed value of 0.55.

The number of turns on each field coil is:

$$N_f = \frac{h_f b_{fc} f_{cf}}{a_{fc}} = \frac{28 \times 40 \times 0.57}{0.159} = 4015$$

The field winding resistance is obtained from equation (4.33):

$$R_f = 2 \times 4015 \times \frac{1.8 \times 10^{-8} \times 0.468}{0.159 \times 10^{-6}} = 426\,\Omega$$

The field current is:

$$I_f = \frac{110}{426} = 0.258\,\text{A}$$

and the corresponding current density is about $1.62\,A/mm^2$; which is quite acceptable.

Next, we compare $I_f N_f$ with the required ampere-turns per pole. The $I_f N_f$ product gives 1035 ampere-turns, and 940 ampere-turns per pole are required. The shunt-field resistance is calculated at 25 °C; the operating temperature will, undoubtedly, be higher than this value. The pole ampere-turns will be less than 1035 at normal operating temperatures.

Finally, we examine the power generated per unit area in relation to the permissible value, between 500 and $750\,W/m^2$. The total area of the cooling surface of the field winding is obtained from equation (4.36):

$$A_{sf} = 2 \times (0.468 + 0.04) \times 0.028 \times 2 = 0.057\,\text{m}^2$$

The power generated per unit area is:

$$\frac{I_f^2 R_f}{A_{sf}} = \frac{(0.258)^2 \times 426}{0.057} = 498\,\text{W/m}^2$$

This design is obviously quite satisfactory.

4.4.5 Design of commutating poles

The width of the interpole (w_{ip}) depends primarily on the width of the commutation zone (w_c). Therefore, we design the commutator and brush gear before designing the interpoles.

Equations (4.39) and (4.40) yield:

$$17.9 \leq w_{ip} \leq 22.3\,\text{mm}$$

The width of the commutation zone for wave-wound motors is given in equation (4.38). In the present case, where there are 2 coil sides per slot and 2 poles, the width of the commutation zone is:

$$w_c = \frac{1}{0.6}(w_b - w_m)$$

Because of the high speed of this machine, we use resin-bonded graphite brushes. The commutator pitch is about 7 mm. A brush 16 mm wide, along the commutator surface, will cover just over 2 commutator segments. The thickness of the insulating spacer may be taken as 0.8 mm. The resulting width of the commutation zone is 24 mm, which is satisfactory. The axial brush length is determined from its current density and the armature current. The current density of resin-bonded graphite is 0.1 A/mm^2, as given in Table 4.1. A brush axial length of 10 mm will be satisfactory. The radial length of the standard brush size is 32 mm.

Let us use the same air-gap for the interpoles as the main poles, i.e. 1.0 mm. The width of the interpole is chosen as 22 mm to allow for the fringing flux at its tips. The interpole axial length is the same as the main pole, or 118 mm.

The armature ampere-turns per pole is 691. we take the interpole ampere-turns 20% higher than the armature mmf per pole or:

$$AT_{ip} = 1.2 \times 691 = 829$$

and the number of interpole turns is:

$$N_{ip} = \frac{829}{15.8} = 52$$

Assuming the current density in the commutating winding equal to 2.2 A/mm^2, the resulting cross-section area of its conductor will be 6.88 mm^2.

The nearest standard wire has a copper diameter of 3 mm and cross-section area of 7.069 mm^2. The diameter of the doubly insulated conductor is 3.14 mm. The copper area of 52 coils is 368 mm^2. Assuming a space factor for the commutating pole coil of 0.5, the required total window area will be 736 mm^2. Therefore, let the average depth of the commutating pole coil to be about 22 mm and its average height 30 mm. The length of the mean turn is:

$$l_{ic} = 2(0.118 + 2 \times 0.03 + 0.022) = 0.4\,\text{m}$$

The total resistance of the commutating field winding is obtained with the aid of equation (4.44) as follows:

$$R_{ip} = \frac{1.8 \times 10^{-8} \times 0.4 \times 52}{7.069 \times 10^{-6}} \times 2 = 0.106\,\Omega$$

The voltage drop in the commutating pole winding is about 1.7 V.

4.4.6 Losses and efficiency

Copper losses

The resistance of the armature winding, shunt field winding and the commutating winding are 0.227 Ω, 426 Ω and 0.106 Ω grespectively. These values are obtained at 25 °C. The corresponding values at 75 °C are:

$$R_a = 0.27\,\Omega \qquad R_f = 506\,\Omega \qquad R_{ip} = 0.126\,\Omega$$

The field current at 75 °C is:

$$I_f = \frac{110}{507} = 0.22\,\text{A}$$

and the armature current at 75 °C is 15.78 A.

The armature copper loss is:

$$W_a = (15.78)^2 \times 0.27 = 67.2\,\text{W}$$

The commutating field winding copper loss is:

$$W_i = (15.78)^2 \times 0.126 = 31.4\,\text{W}$$

The shunt-field winding copper loss is:

$$W_f = 0.22^2 \times 507 = 24.5\,\text{W}$$

The brush contact voltage drop for the resin-bonded graphite brushes is assumed to be 2 V. The brush contact I^2R loss is:

$$W_b = 2 \times 15.78 = 31.7\,\text{W}$$

The total copper loss is 154.8 W.

Core losses

The weight of the armature teeth and core are obtained with the aid of equations (4.49) and (4.50) as follows:

$$G_t = 25 \times 0.0045 \times 0.95 \times 0.118 \times 0.015 \times 7750 = 1.47\,\text{kg}$$

$$G_c = \frac{\pi}{4}[(0.095 - 2 \times 0.015)^2 - (0.025)^2]0.95 \times 0.118 \times 7750 = 2.46\,\text{kg}$$

Specific iron loss values at 50 Hz and at 1.53 T and 1.5 T are obtained for the teeth and armature core, respectively, from the lamination's core loss curve. The iron loss values, in W/kg at 50 Hz, are 4.2 and 4.1 for the teeth and armature core respectively. At 45 Hz, these values are 3.8 W/kg and 3.7 W/kg. The armature teeth and core iron losses are:

$$1.47 \times 3.8 + 2.46 \times 3.7 = 14.7\,\text{W}$$

The total iron losses are estimated by multiplying the armature teeth and core losses by 2.5. Therefore, the total core loss is 37 W.

Brush friction loss

The friction loss for graphite brushes is determined in terms of the total brush contact area and the commutator peripheral speed as discussed in Section 4.3.12. The commutator speed is 8 m/s and the total brush contact area is 320 mm^2. Therefore

$$W_{bf} = 2440 \times 320 \times 10^{-6} \times 8 = 6.3\,\text{W}$$

Bearing friction and windage losses

In the absence of test data on similar machines, we estimate these losses as $v_a/50\%$ of the output power. The armature peripheral speed is about 13.4 m/s. The bearing friction and windage losses are estimated to be 4 W.

Efficiency

The total losses in the motor are:

$$\text{Total losses} = 154.8 + 37 + 6.3 + 4 = 202.1\,\text{W}$$

and the efficiency is:

$$\eta = \frac{1500}{1500 + 202.1} = 88.1\%$$

5

INDUCTION MOTOR DESIGN

5.1 INTRODUCTION

Induction machines are usually operated as motors and they provide the most common form of drive for industrial and domestic applications. The popularity of the induction motor stems from its simple construction which makes it less expensive than a d.c. motor of the same rating. Also, induction motors require practically no maintenance.

The main disadvantages of induction motors are their high starting current and their low power factor at light loads. However, for many applications, their advantages outweigh their disadvantages.

The speed of an induction motor is very nearly constant. Traditionally, induction motors were used in applications requiring constant speed operation. Recently, with advances in the area of electronic power conversion, induction motors have become widely used in variable-speed drives for various industrial and processing applications.

5.2 CONSTRUCTIONAL DETAILS

5.2.1 Conventional motors

The operation of an induction motor requires two windings, a primary, which is excited by an alternating current source, and a secondary, which is short-circuited. In the conventional design, the primary winding is housed on the stator while the rotor carries the secondary winding, but this not essential to the concept.

The primary always utilises a distributed winding, wound to give the required number of poles and phases. There are two types of secondary windings giving rise to two types of induction motors; the slip-ring (or the wound-rotor) and the squirrel cage.

The rotor of the slip-ring motor carries an insulated polyphase winding, wound to give the same number of poles as the stator winding. The winding leads are connected to slip-rings mounted on the shaft. Carbon brushes riding on the slip

Figure 5.1 Cutaway three-phase squirrel-cage induction motor.

rings allow insertion of external resistances into the rotor circuit to improve starting characteristics. During normal operation the slip-rings are shorted together. Generally, wound-rotors are only used in large motors.

The squirrel-cage winding consists of solid bars of copper or aluminium joined at each end by an end-ring of the same material. Rotors of small machines usually utilise die-cast aluminium bars and end-rings. Figure 5.1 shows a cutaway view of a three-phase squirrel-cage induction motor.

Both the stator and the rotor of an induction motor utilise slotted laminations. The laminations used in industrial motors are either 0.35 or 0.5 mm thick non-oriented low-loss electrical steel. The stator and rotor of domestic appliances fractional horsepower motors are generally built from silicon-free materials-and the lamination are usually 0.65 mm thick.

Motors having an outside diameter of the stator core of up to 1metre use one-piece core laminations. The stator and rotor are punched from the same sheet. For small motors, partly closed slots are used and the teeth, not the slots, have parallel sides, as illustrated in Figure 5.2 [5.1].

The stator laminations are assembled into a core pack. The stack is pressed and the laminations are then either welded or riveted together. In the case of a welded stator core, the weld is applied directly to the laminations in groves specially punched for this purpose. Figure 5.3 shows a riveted stator core assembly.

Rotor laminations of wound-rotor motors are usually assembled directly on the shaft, keyed to it and pressed together with the aid of washers or thrust rings. In

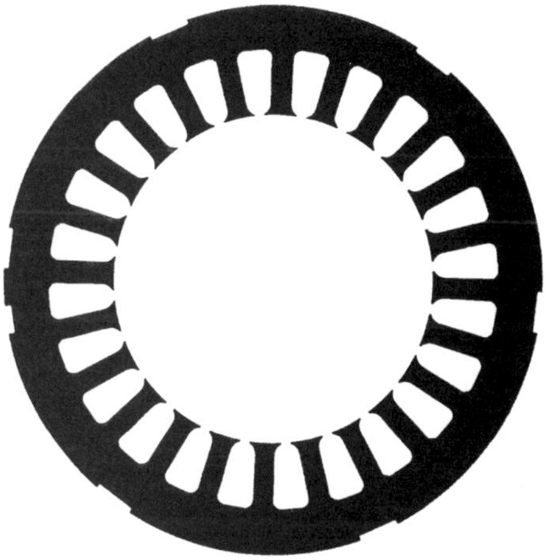

Figure 5.2 Partly closed tapered slots used for small induction motors.

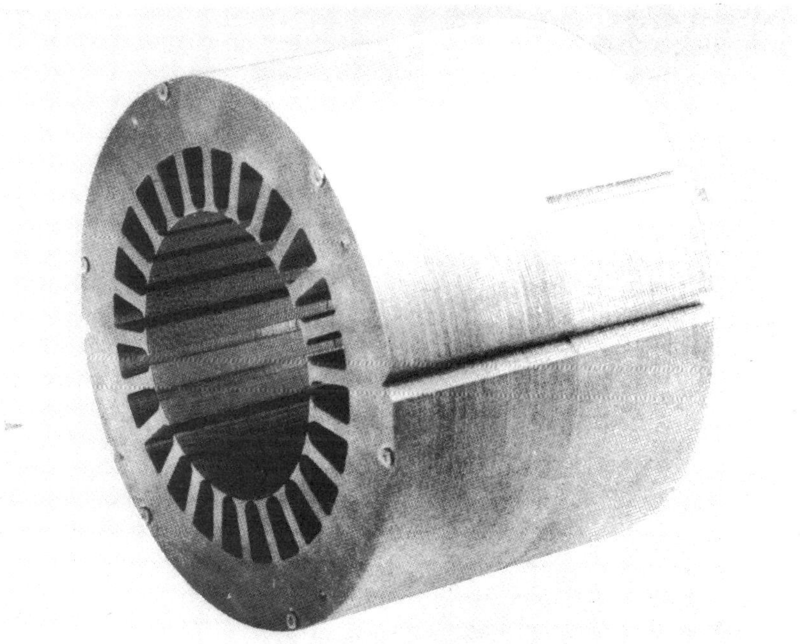

Figure 5.3 A riveted stator core assembly.

the case of squirrel-cage rotors, the laminations are stacked in a core mould into which molten aluminium is forced under pressure. With this production method, the rotor bars, end-rings, and cooling fins are cast at the same time and the rotor is manufactured in a single process. In practice, the aluminium is alloyed with

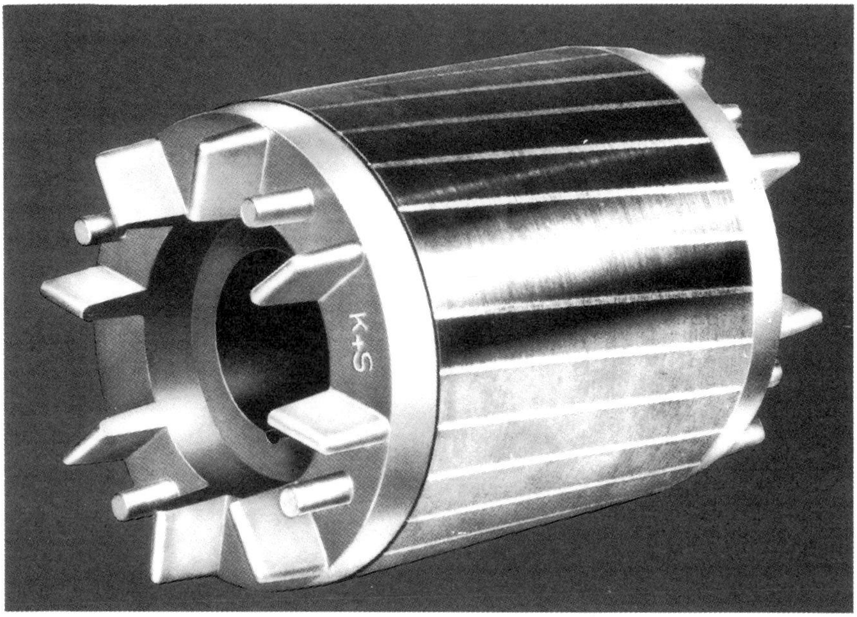

Figure 5.4 A diecast squirrel-cage rotor.

silicon (between 6 and 12%) as pure aluminium does not cast well. Figure 5.4 shows a squirrel-cage rotor.

In some designs the squirrel-cage winding utilises round or rectangular copper bars cut to match the shape of rotor slots. The bars are driven into rotor slots, projecting a short distanceon each end of the core. The bar ends are then soldered, brazed or welded to the end-rings. This manufacturing method is obviously more expensive than die-casting. It may be noted that die-casting of a copper-cage winding is possible, but it requires a much higher casting temperature than that required for aluminium casting.

5.2.2 Shape of rotor slots

The starting and running characteristics of a squirrel-cage induction motor can be tailored to suit a given application by proper shaping of rotor bars and by changing their position with respect to the air-gap magnetic field. Figure 5.5 illustrates different designs of cage windings.

For example, using large bars results in a low rotor resistance, and when the cage winding is positioned near the rotor surface the secondary leakage reactance is reduced, as the cage winding becomes more closely coupled to the stator winding. Such a motor would have low slip and high efficiency at full load and high pull-out torque, but it would also have low starting torque and high starting current. This design is designated as "Design A" by NEMA and by various other national and international standards. In design F, Figure 5.5, the rotor bars have the same cross-section area as design A but the bars are buried deep in the rotor, and this

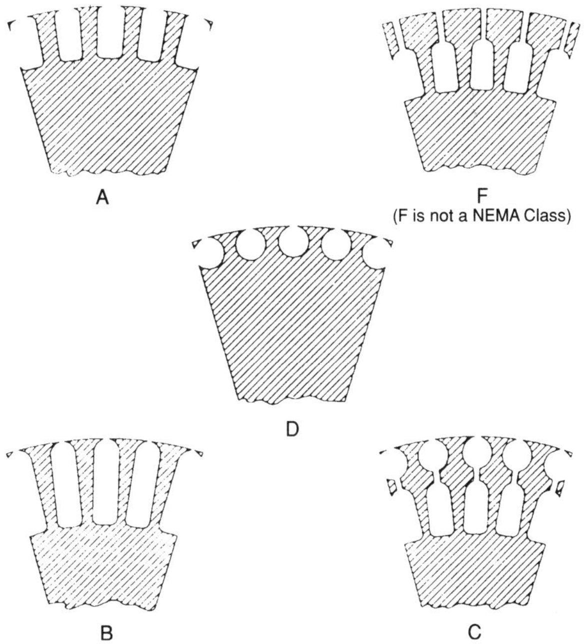

Figure 5.5 Squirrel-cage rotor-bar cross section to produce NEMA design-class characteristics.

increases the leakage reactance. The rotor resistance is still low and the full load characteristics would remain largely as those obtained from design A. However, the larger rotor leakage reactance in design F limits the starting current, but it also makes the starting torque low and reduces the pull-out torque. It should be noted that design F is no longer a NEMA class.

In design D, Figure 5.5, the rotor leakage would be similar to that of design A, as the cage winding position with respect to the stator is the same in both cases. However, the bar cross-section area is reduced in design D, and this increases the rotor resistance. Therefore, a NEMA design D rotor has high starting torque, low starting current, and high slip and efficiency at full load.

The desired characteristics of designs F and D may be combined in the NEMA designs B and C, Figure 5.5. Design B results in the so called deep-bar rotor, while design C is termed a double-cage rotor.

Bearing in mind that the depth of flux penetration into the rotor is inversely proportional to the square root of rotor frequency, the theory of the deep bar rotor may be explained as follows: at starting, the rotor frequency is high, and the rotor flux links only the top sections of the rotor bar. As the rotor starts to rotate, its frequency decreases, reaching a minimum value at full load, the depth of flux penetration increases and the effective bar section increases. Therefore, the effective rotor resistance is high at starting, giving good starting characteristics, and is low at full load, giving low slip and high efficiency.

In a similar way, the high resistance, low inductance, outer cage of design C becomes the only effective section during starting and as the motor accelerates the

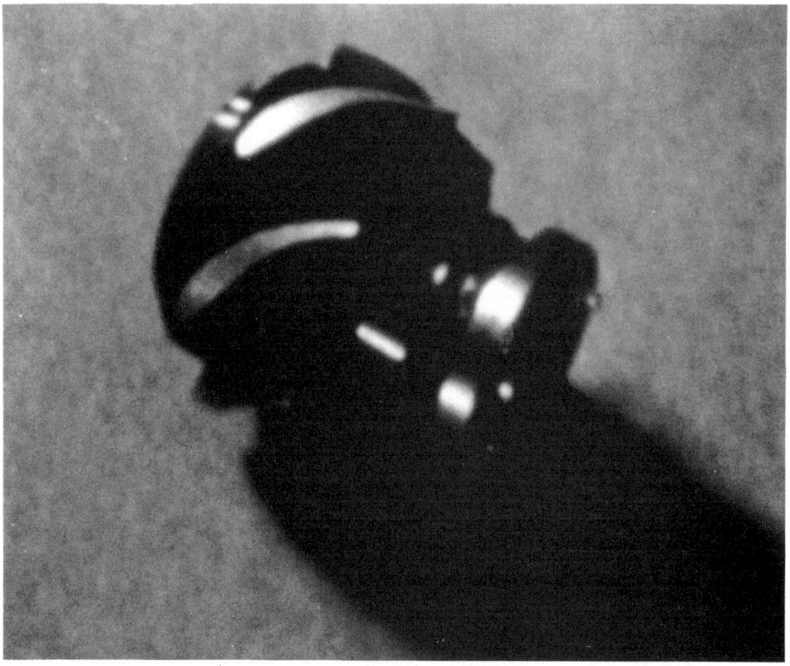

Figure 5.6 An outer-rotor induction motor.

flux penetrates deeper into the rotor and links the low resistance inner cage winding. At full load, although the two windings will carry current and produce torques, the inner cage develops, typically, 90% of the torque.The torque, speed and current requirements of the designs shown in Figure 5.5 (except for design F which is no longer listed as a standard) are given in NEMA Standard MG-1. These are summarised in reference [5.2, p. 274].

5.2.3 Some special designs

Outer rotor motors

In some special designs, the position of the primary and secondary windings is interchanged; providing the so called "inside out or outer-rotor" motor. For example, in some high speed, 400 Hz, fan-drive applications where the motor and fan size are very small, it is advantageous to utilise an outer-rotor design and to make the fan blades an integral part of the (outer) rotor. One such arrangement is illustrated in Figure 5.6.

Solid rotor motors

The characteristics of motors utilising plain unlaminated steel rotors are quite different from those offered by the conventional squirrel-cage design. This is due to the impedance of the solid rotor being inversely proportional to the product of

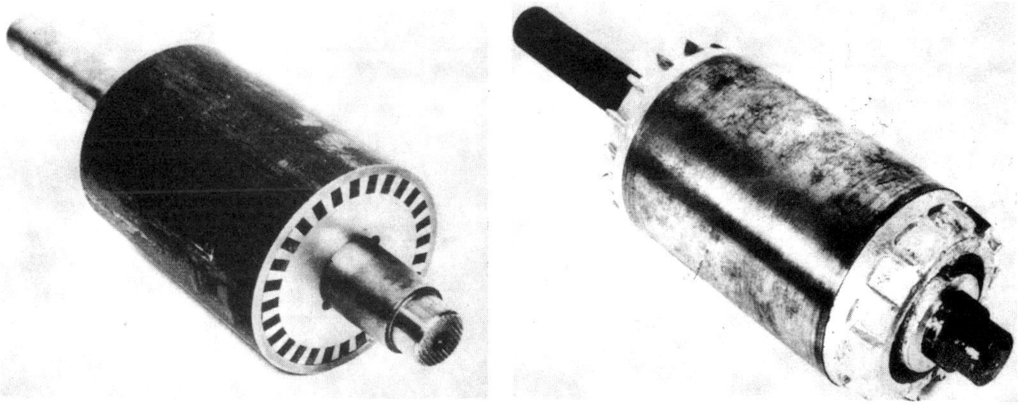

Figure 5.7 A composite-rotor before and after cast of cage winding.

air-gap flux and slip [5.3].

An unlaminated rotor offers an inherent high resistance characteristic as well as having advantageous mechanical and thermal attributes. This type of construction was investigated for variable-voltage fan drives [5.4]. In such applications the effective high resistance of the solid rotor yields reduced primary current and stator heating at low speeds. However, the slip at rated voltage tends to be higher than that of a cage rotor and this results in a reduction of both fan speed and output. Techniques are available for relieving this problem, such as fitting copper end-plates or machining deep radial slits in the rotor surface [5.5]. Such techniques, however, are not considered to be entirely practical.

Composite-rotor motors

Developed by Chalmers and Hamdi [5.6, 5.7], this design combines the respective advantageous properties of a low resistance cage winding and a solid rotor (low slip and high output at full load and low current and high torque at starting, respectively). This construction has two different regions; the inner is a low resistance normal cage, while the outer is a steel shell [5.6]. The steel shell was freeze fitted on to the laminated core and machined after casting of the cage winding to give the desired air-gap. Figure 5.7 shows a typical composite rotor [5.6].

The thickness of the outer steel shell can be so chosen that it carries the magnetic flux up to a certain voltage level. At higher voltages, the flux penetrates into the inner cage region which yields a reduced effective resistance. The resulting characteristic is in between that of a solid rotor and a squirrel-cage motor; the steel shell/cage combination can be optimised to give the desired characteristics.

5.3 DESIGN CALCULATIONS

The output equation of a.c. machines is derived in Chapter 3 and is reproduced here for convenience:

Table 5.1 Efficiency and power factor of three-phase, 4-pole squirrel-cage induction motors

Output (kW)	Efficiency	Power factor
0.25	0.68	0.7
0.75	0.72	0.75
2.2	0.81	0.82
3.7	0.83	0.84
7.5	0.86	0.87
15.0	0.88	0.89
37.0	0.9	0.9

$$Q = C_0 D^2 L n_s \tag{5.1}$$

and,

$$C_0 = 1.11\pi^2 B_{av} ac K_w \times 10^{-3} \tag{5.2}$$

where Q is the motor's kVA input, B_{av} is the average flux density in the air-gap, T, ac is the specific electric loading, ampere conductor per metre, D and $L$are the air-gap diameter and core length, m; respectively, K_w is the winding factor and n_s is the synchronous speed, rev/s.

The kVA input (Q) is determined in terms of the rated output power (P_o), efficiency (η) and power factor as follows:

$$Q(\text{kVA}) = \frac{P_o(\text{kW})}{\eta \cos \phi} \tag{5.3}$$

The winding factor (K_w) is:

$$K_w = K_d K_c \tag{5.4}$$

where K_d and K_c are the primary winding distribution and the chording (or pitch) factor, respectively.

Initially, the distribution factor may be taken as 0.955 for three-phase motors (corresponding to 3 slots per pole per phase) and the pitch factor may be assumed unity. The winding factor will be determined when the stator (primary) winding is designed. Initial values of efficiency and power factor may be taken from Table 5.1. These are typical values for supply frequency motors.

5.3.1 Air-gap flux density

In an induction motor, the magnetising current required to establish the flux in the magnetic circuit is supplied through the primary winding. It lags the applied voltage by 90° and must be minimised in order to obtain good operating characteristics. The magnetising current is predominately determined by the mmf required to pass the flux across the air-gap between rotor and stator, even with the smallest possible air-gap. The gap mmf is directly proportional to the radial length of and the flux density in the air-gap. Therefore, to avoid excessive magnetising currents induction motors are designed for moderate values of flux densities.

As well as affecting the magnetising current and power factor, the value of the air-gap flux density determines the teeth density and the motor's overload capacity. High flux densities in the teeth give rise to high core losses and high magnetising currents. Increasing B_{av}, and the flux per pole, results in a smaller number of turns per phase being required for the stator winding and, consequently, a decrease in the leakage reactance. With low value of leakage reactance, and for a given terminal voltage, the diameter of the motor's circle diagram increases [5.7, p. 293]. This increases the maximum power which the motor can deliver, or, in other words, its overload capacity. Thus, the higher the design value of B_{av} the higher the overload capacity will be. However, the air-gap flux density should be so chosen that there is no saturation in any part of the magnetic circuit. In most cases, this provides a motor with an overload capacity of about twice the rated output (or 100%).

As the number of poles increases (or as the speed decreases) it becomes more difficult to get a 100% overload capacity with a satisfactory power factor. In such a case a compromise between these two characteristics has to be made.

Satisfactory designs of supply frequency induction motors of ratings up to 22 kW are generally obtained with air-gap flux density of between 0.35 and 0.6 T; the high values are for the larger motors. Smaller values of B_{av} may be used for the design of 400 Hz motors.

5.3.2 Specific electric loading

The value of ampere-conductors per metre of stator circumference at the air-gap (*ac*) depends on the motor size, the type of enclosure and ventilation and the desired overload capacity (or the permissible leakage reactance).

A high value of *ac* results in high copper losses and high temperature rise. The ventilation system employed must be so designed that the temperature rise does not exceed the maximum allowable value determined by the class of insulation used.

A high value of ampere-conductors per metre would result in a large number of conductors being required, and this increases the leakage reactance of the machine. For a given terminal voltage, the overload capacity of the motor will be reduced.

In case of small motors, the specific electric loading (*ac*) is usually between 8000 and 25,000 ampere-conductors per metre. Higher values are used for motors designed to operate immersed in liquid coolant.

5.3.3 Number of poles

The number of poles is readily found in terms of the synchronous speed and supply frequency as follows:

$$p = \frac{2f}{n_s} \tag{5.5}$$

where the synchronous speed n_s is in rev/s.

5.3.4 Main dimensions

After selecting initial values for specific electric and magnetic loadings, power factor and efficiency, the output coefficient and the product D^2L are calculated from equations (5.2) and (5.1); respectively. The factors which needs to be considered when separating D and L for induction motors are discussed in chapter 3. In brief, the ratio of the core length to pole pitch may be selected as follows:

L/τ between 1.5 and 2.0:	gives minimum cost
between 1.0 and 1.5:	gives good power factor
equals to 1.5:	gives good efficiency
equals to 1.0:	gives a balanced design

For small motors, high values of L/τ result in small diameters which may not be able to accommodate even a small number of slots. A value of L/τ as low as 0.6 may be used for 2-pole machines having a rating of a few kilowatts and for some fractional-horsepower motors. Therefore, in the general case:

$$L = \frac{\pi D}{p} \times (0.6 \text{ to } 2.0) \tag{5.6}$$

Rotor peripheral speeds of up to 75 m/s do not require any special mechanical construction features. This value is unlikely to be exceeded in the motors treated in this book.

5.3.5 Air-gap length

The ampere-turns required to pass the flux in the air-gap are directly proportional to its radial length and the flux density value. Even with low air-gap densities and short air-gap length, the air-gap ampere-turns are larger than the mmf required to pass the flux in the remainder of the magnetic circuit. In order to minimise the magnetising current and to obtain satisfactory performance the air-gap should be as small as the mechanical construction will permit.

There are, however, some advantages resulting from using large air-gaps. Pulsation losses and magnetic noise, which are due to reluctance variation of the air-gap, decrease as the air-gap is increased. Unbalanced magnetic pull is more likely to occur with small gaps. Large air-gaps improve the ventilation conditions in fan-cooled motors. In most cases, the requirement of low power factor limit the air-gap radial length to the minimum mechanical clearance.

The approximate minimum air-gap length, l_g in millimetres, can be determined by the following empirical formula:

$$l_g = 3.06 - \frac{6560}{D + 2280} \tag{5.7}$$

where D is the stator air-gap diameter, mm.

5.3.6 Stator design

5.3.6.1 *Stator slots*

Small motors utilise partly closed tapered slots with parallel-sided teeth. The need for partly closed slots arise because of the small value of air-gap (less than 1 mm); this arrangement reduces the flux pulsations. The use of tapered slots yields maximum slot area for a given tooth density.

There are no definitive rules to guide the designer in selecting the number of stator slots. However, a large number of slots require more slot insulation and large numbers of coils to wind, insulate and install, and this leads to a higher manufacturing cost. On the other hand, the slot leakage flux, and its contribution to the motor's leakage reactance, is inversely proportional to the number of slots per pole per phase. Therefore, the number of slots per pole per phase should not be less than two otherwise the leakage reactance becomes high.

To keep the manufacturing cost as low as possible, the number of stator slots should be so chosen that the maximum number of poles, phases and voltages combinations may be realised within a given frame size.

Standard induction motors are usually designed for dual voltage operation. This can be achieved by using, at least, a two circuit winding. The winding circuits will be connected in series for operation at a given voltage and in parallel for operation at half the given voltage. The number of parallel circuits is very much limited with fractional slot windings because these windings do not repeat every pole, unlike the windings with integral slots per pole per phase. Thus, standard induction motors are generally designed with an integral number of slots per pole per phase not less than 2.

As well as meeting the requirements of the number of poles and phases, the number of stator slots must also be selected so that the resulting number of conductors per slot is an integer for a single layer winding, and an even-integer for a double layer winding. However, short pitching the winding provides a means of effectively changing the number of conductors as is discussed in the following section.

5.3.6.2 *Stator winding*

The stator winding may be either of the single layer type (with one coil side per slot), or of the double layer type. Single layer mush (or basket) windings are the most commonly used type for small induction motors. Single layer windings of the concentric type and double layer windings are used for some industrial motors.

In the case of the mush windings, all the coils have the same span and the coil ends belonging to coils situated in consecutive slots cross one another, i.e. they proceed to left and right alternately. There is only one coil group per phase per pole-pair and, therefore, the maximum number of parallel circuits is equal to half the number of poles. The coil span, in terms of number of slots, of mush windings must be odd for three-phase motors and even for two-phase motors.

A sine wave flux distribution may be assumed for induction motors as the distributed stator winding produces an air-gap flux waveform which is very nearly sinusoidal. The voltage induced in the armature winding is [5.2, p. 69]:

$$E_s = 4.44 f \phi T_s K_{ws} \tag{5.8}$$

where E_s is the voltage induced in the stator (primary) winding per phase, V, f is the supply frequency, Hz, ϕ is the flux per pole, Wb, T_s is the number of stator winding turns in series per phase and K_{ws} is the stator winding factor.

The induced voltage per phase may be taken equal to the terminal phase voltage minus the product of the magnetising current and stator leakage. Initially, E_s may be taken as 0.97 times the terminal phase voltage.

The flux per pole is determined in terms of the average flux density in the air-gap (B_{av}), pole pitch (τ) and the nett core iron length (L_i) as follows:

$$\phi = B_{av} \tau L_i \tag{5.9}$$

where ϕ is in Weber, τ and L_i are in metres and B_{av} is in Tesla.

The number of stator turns per phase is:

$$T_s = \frac{E_s}{4.44 f \phi K_{ws}} \tag{5.10}$$

The total number of conductors in the stator winding is $6T_s$ for three-phase motors. It is inversely proportional to the winding factor. The distribution factor (K_d) is fully determined by the number of slots per pole. The chording factor is defined as the sine of half the angle spanned by the coil. The chording factor for a coil spanning 180 electrical degrees will be 1.0 and the emf induced in the coil will be at the maximum possible value. If there are 9 slots per pole (giving a slot angle of 20 electrical degrees) and one coil side is in slot number 1 and the other coil side is in slot number 8, the coil span will be 7 slots or 140 electrical degrees. The chording factor will be:

$$K_c = \sin \frac{140}{2} = 0.94 \tag{5.11}$$

When the armature winding required to give a good design yields a fraction or an odd number of conductors per slot, we choose the next larger integer or even number, for a single layer or a double layer winding, respectively, with short-pitched (or chorded) coils.

Additional advantages of using short-pitched coils are that the mean length of armature turn is reduced and the induced voltage waveform improves.

The stator phase current is obtained as follows:

$$I_s = \frac{P_o \times 10^3}{3V_{ph} \cos \phi\eta} \tag{5.12}$$

The conductor current, I_{zs}, is obtained by dividing the motor phase current by the number of parallel armature circuits. The copper losses in any winding vary directly with the squared value of its current density. The temperature rise depends on the losses for a given construction. The stator current density must be chosen so that the required efficiency is obtained without exceeding the temperature rise limits. The current density in the stator winding of a standard induction motor may be assumed between 3 and 6 A/mm^2; depending on the duty type and the ventilation method employed. The cross-section area of the stator conductor is:

$$a_s = \frac{I_{zs}}{\delta_s} \quad \text{mm}^2 \tag{5.13}$$

where δ_s is the current density in the stator conductors, A/mm^2.

Generally, the requirements of small induction motors are met by round conductors. The wire size may be selected from the standard wire table given in Appendix 1.

For a single-layer winding the approximate length of the mean turn (L_{mts}) may be determined from the following empirical formula:

$$L_{mts} = 2L + 2.3\tau \tag{5.14}$$

where the core length L and the pole pitch τ are in metres.

For a double-layer winding the length of the mean turn will be between 10 and 25 mm greater than the value given in equation (5.14).

When the number of conductors per slot and the conductor cross-section area are known, an approximate slot area can be obtained by assuming a value of slot space factor.

Small induction motors utilise semiclosed slots; the slot insulation is inserted into the slots, the stator coils are automatically wound and then placed in the slot. This manufacturing method results in a typical value of slot space factor of about 0.4, generally not higher than 0.5. Therefore:

$$A_{ss} = \frac{Z_{ss} a_s}{0.4} \tag{5.15}$$

where Z_{ss} is the number of conductors per slot. The stator slot area A_{ss} and the conductor cross-section area a_s are in millimetres squared.

Deep slots result in higher slot leakage and, generally, the ratio of slot depth to width should be between 3 and 6. However, the dimensions of the slot determine the flux density in the teeth. The slot should be so proportioned that resulting maximum tooth flux density is about 1.7 T.

The stator resistance per phase is:

$$r_s = \frac{\rho T_s L_{mts}}{a_s \times 10^{-6}} \tag{5.16}$$

where a_s is in mm^2 and ρ is in Ωm.

5.3.6.3 Stator teeth and core

The core losses and magnetising mmf vary directly with the tooth flux density. Therefore, a high flux density in the teeth is not desirable. As stated above, the tooth density should not exceed 1.7 T. The minimum stator tooth width, W_{ts} in metres, is given by:

$$W_{ts} = \frac{p\phi/S_s}{1.7L_i}$$

where S_s is the number of stator slots and L_i is the core iron length.

The average flux density in the stator core should not exceed 1.5 T. Bearing in mind that half the pole flux passes through the core, the minimum core depth, d_{cs} in metres, is given by:

$$d_{cs} = \frac{\phi/2}{1.5L_i} \tag{5.17}$$

The outside stator diameter, D_o in metres, is:

$$D_o = D + 2d_{ss} + 2d_{cs} \tag{5.18}$$

where d_{ss} is the depth of stator slot in metres.

5.3.7 Design of a squirrel-cage rotor

The theory of polyphase induction motors and the development of its current, torque and power relationships are based on the assumptions that the synchronously rotating mmf produced by stator winding excitation is sinusoidally distributed in space over all the principal poles, and that the rotor mmf due to the slip frequency induced currents is similarly distributed over all the induced poles.

In practice, however, the air-gap flux distribution contains a number of harmonics. These are produced due to: the concentration of mmf in a finite number of slots (the winding harmonics), and the fact that the air-gap is not smooth because of the stator and rotor slot openings (the slot harmonics). Additionally, harmonics may be produced due to saturation and irregularity in the air-gap length.

The harmonic fluxes may be regarded as additional sets of poles, some of them travelling forward (in the direction of the fundamental main field), others in the backward direction, but all travelling at sub-synchronous speeds. The squirrel-cage rotor reacts to each harmonic flux in the same manner as to the fundamental and,

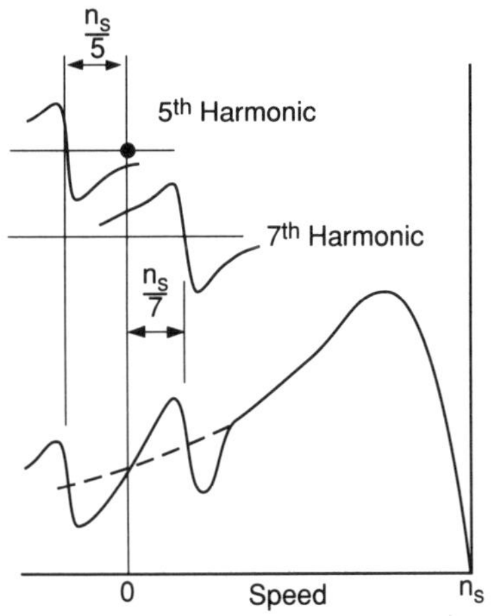

Figure 5.8 Dips in the torque–speed characteristics caused by 5th and 7th winding harmonics.

thus, produces a complete "parasitic" speed–torque characteristic for each harmonic. Consequently, the speed–torque characteristic of the motor is not the smooth calculated curve but has irregularities, dips and cusps, at large slips.These may result in a motor that tends to lock at standstill and fails to start, tends to crawl (stably run) at some sub-synchronous speed, or may result in excessive noise during the run-up period.

It should be noted that slip-ring motors are less likely to have these troubles because their rotors are wound for a specific number of poles and, therefore, less sensitive to harmonics and, in any case, it is possible to increase the rotor resistance so that the motor starts on its pull-out torque.

5.3.7.1 *Number of rotor slots*

Winding harmonics A three-phase winding carrying sinusoidal currents produces harmonics of the order n given by:

$$n = 6N \pm 1 \tag{5.19}$$

where N is an integer. The direction of rotation of the harmonic is with or against the direction of rotation depending on the sign (+ gives the forward rotating harmonics, and − yields the backwards ones). The number of poles of the nth harmonic is n times the number of poles of the fundamental and, therefore, its synchronous speed is $1/n$ times the synchronous speed of the fundamental.

A three-phase winding will produce a forward rotating 7th harmonic and a backward rotating 5th (for $N = 1$). Each of these two harmonics generate rotor emfs, currents and torques in a similar way to those generated by the fundamental. The torque–speed characteristics developed by the 5th and 7th harmonics are of the same general shape as that of the fundamental but with synchronous speeds $1/5$ (backward) and $1/7$ (forward) times the fundamental synchronous speed, as shown in Figure 5.8. The resultant torque/speed curve shows a marked dip at large slips, and, with certain stator and rotor slot combinations, the dip due to the 7th harmonic may become very pronounced. Assuming a constant torque load, the motor torque may fall below that of the load and, if that occurs, the motor will fail to accelerate to its full speed and continues to run at a speed just less than the 1/7 the synchronous speed. This is called crawling.

Slot harmonics Tooth ripples or slot harmonics are produced by the effect of reactance variations as the slotted rotor moves with respect to the slotted stator. A complete cycle of this tooth ripple takes place in the time required for the pole to move through an angle equal to one slot pitch. Thus, if there are q slots per pole per phase, and m phases, there would be $2mq$ cycles of the tooth ripple for one complete cycle of fundamental frequency f. Therefore, the frequency of the tooth ripple flux, f_t, is:

$$f_t = 2mq \tag{5.20}$$

The tooth ripple flux of frequency f_t can be shown to produce tooth ripple emf with two frequencies f_{t1} and f_{t2} as follows:

$$f_{t1} = f_t + f = f(2mq + 1), \qquad f_{t2} = f_t - f = f(2mq - 1) \tag{5.21}$$

So far, only the fundamental of slot reluctance variation has been considered. Further ripples may occur due to harmonics. In general, the slot harmonic emfs of frequencies:

$$(Af_t \pm 1), \qquad \text{where } A \text{ is an integer} \tag{5.22}$$

are produced. For a three-phase winding, the order of slot harmonic n is given by:

$$n = 6Aq \pm 1 \tag{5.23}$$

Considering a 4-pole, 36 slot machine ($q = 3$), the 17th (backward) and 19th (forward) harmonics are produced due to slotting. Thus, the dips in the torque–speed characteristic would be produced at $+1/19$ and $-1/17$ of synchronous speed as shown in Figure 5.9.

In order to illustrate how the effect of dips in the torque–speed characteristic may be augmented by rotor slotting, consider the case where a 76-slot (or bar) rotor is used with this stator. In this case, there would be one rotor bar corresponding to every 19th harmonic pole. Thus, the 19th harmonic torques would be very large resulting in considerable vibrations at starting.

So far only the stator slot harmonics are considered. The rotor also produces slot harmonics. If the stator and rotor harmonics of the same order are present, and if their speeds happen to coincide, they will lock together giving rise to synchronous torque. If such a torque is sufficiently large, the motor will crawl at a constant sub-synchronous speed. For a three-phase motor with p poles, the order of the stator and rotor field harmonics due to slotting, respectively, will be:

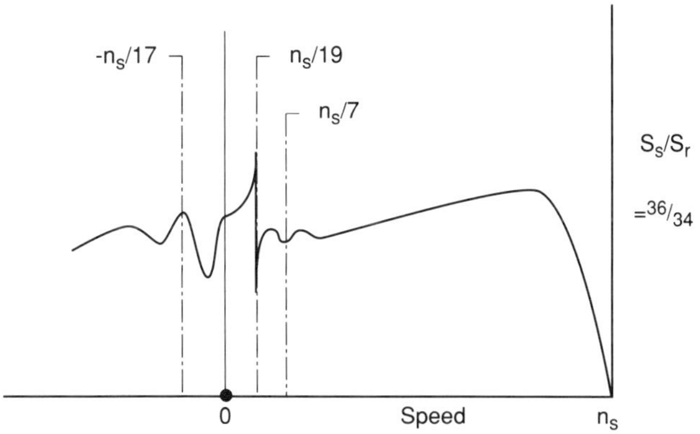

Figure 5.9 Dips in the torque–speed characteristics caused by 17th and 19th slot harmonics.

$$n = 6q \pm 1 = \left(\frac{2S_s}{p}\right) \pm 1 \qquad \text{(for } A = 1\text{)} \tag{5.24}$$

These harmonics revolve at a speed $1/n$ of synchronous speed with respect to stator, and;

$$n' = \left(\frac{2S_r}{p}\right) \pm 1 \tag{5.25}$$

and they rotate at a speed $1/n'$ with respect to the rotor. The speeds will be equal when:

$$\left(\frac{2S_s}{p}\right) \pm 1 = \left(\frac{2S_r}{p}\right) \pm 1 \tag{5.26}$$

One of the possibilities for this to happen is, obviously, when number of stator slots is equal to number of rotor slots. The other possibility is when $S_r - S_s = p$. These situations must be avoided.

Rules for selecting stator–rotor slot combinations The number of stator slots must be an integer and must be such a number that a balanced winding can be accommodated. The total number of slots will always be satisfactory for a given number of poles and phases when the numerator of the fraction (slots/poles) reduced to its lowest terms is a multiple of the number of phases. Similar considerations apply to the selection of rotor slots for slip-ring induction motors, whereas the squirrel-cage rotor always has a balanced winding irrespective of the number of slots. The discussion above shows that the windability is not the only consideration in the selection of the number of stator and rotor slots.

There is hardly any slot combination that will yield a perfectly smooth speed–torque characteristic. A judicious selection of the stator–rotor slot combination, however, does help to alleviate the effect of the most serious harmonics on the performance of the machine. The following general rules should be followed:

(1) As stated above, the number of rotor slots should never be equal to stator slots, but must be either larger or smaller. Satisfactory performance is obtained when the number of rotor slots is 15 to 30% greater or smaller than the number of stator slots.

(2) To avoid synchronous cusps, the difference between stator slots and rotor slots should not be equal to p, $2p$ or $5p$.

(3) The difference between the number of stator and rotor slots should not be equal to $3p$ or any multiple of $3p$ for a three-phase motor in order to avoid magnetic locking.

(4) The difference between number of stator slots and rotor slots should not be equal to 1, 2, $(p + 1)$ or $(p + 2)$ in order to avoid noise and vibrations.

Summary $S_s - S_r$ should not be equal to:

$$0, \pm p, \pm 2p, \pm 3p, \pm 5p$$

$$\pm 1, \pm 2, \pm(p \pm 1), \pm(p \pm 2)$$

Notwithstanding the above rules, there are some slot combinations which have been proved satisfactory in practice, and, therefore, should be used whenever possible. Some of these combinations for industrial 50 Hz motors are:

2 poles	18/23	36/22	36/24	36/28
	36/32	42/32	48/38	54/38
4 poles	24/33	24/17	48/40	
	48/60	60/42	60/52	
6 poles	36/44	36/40	48/44	54/38
	72/54	72/56	72/58	
8 poles	72/52	72/60		

The most desirable stator and rotor slot combinations are not always possible for motors with small diameters. In such cases, magnetic locking, noise and cusps in the speed–torque characteristics can be reduced or even eliminated entirely by skewing either stator or rotor slots and short pitching the winding. The skew is approximately one stator or rotor slot, whichever is larger.

For 400 Hz aircraft motors, the following combinations have been used successfully:

2 poles	12/16	18/23			
4 poles	12/10	12/14	18/22	24/15	24/30
	36/30				

6 poles 12/21 18/22 36/30 36/44

8 poles 36/44

Some of these combinations are obviously unsatisfactory, based on the discussion above, but the shape of the torque–speed characteristic for aircraft applications is seldom critical, and almost all these machines utilise skewed laminations and short-pitched windings.

5.3.7.2 Rotor bars and end-rings

The total rotor ampere-turns may be assumed between 0.85 and 0.9 times the total stator ampere-turns. No insulation is used between the cage bars and rotor core, and the rotor may be worked at a much higher current density than the stator. Additionally, the mean length of rotor turn is shorter than the stator and ventilation is better. The current density in the rotor bars may be taken no greater than twice that in the stator winding. Thus, the ratio of the total rotor conductor section to the total stator copper section is, approximately:

$$\frac{A_{cr}}{A_{cs}} = 0.9 \times \tfrac{1}{2} = 0.45$$

where A_{cr} and A_{cs} are the total conductor section areas for the rotor and stator, respectively.

However, the total rotor conductor area must be selected in relation to the length of the bar and end-ring section so that the proper rotor resistance can be obtained to meet starting torque and running performance requirements. It is generally between 50 and 80% of the total stator copper section.

The cross-section area of each bar is obtained in terms of the total conductor area in the rotor (A_{cr}) and the number of rotor bars, or rotor slots (S_r), as follows:

$$a_b = \frac{A_{cr}}{S_r}$$

No insulation is used between bars and rotor core and a clearance of between 0.15 mm and 0.4 mm may be assumed in calculating the rotor slot area.

The total resistance of the squirrel-cage bars is:

$$r_b = S_r \frac{\rho L_b}{a_b} \tag{5.27}$$

where L_b is the axial bar length.

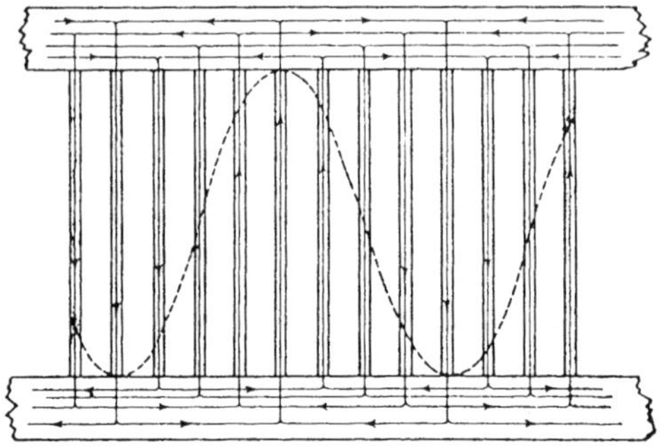

Figure 5.10 Current distribution in a cage winding.

The distribution of the current in the bars and end-rings of a squirrel-cage winding is shown in Figure 5.10. It is shown that the current in each bar divides in the end-ring, one half returning through a bar a pole pitch to the right and the other half through a bar a pole pitch to the left.

It is shown in Figure 5.10 that at points where the current is maximum in the bars; current is zero in the end-rings, the current in the end-rings is maximum where the bar current is zero. A sinusoidal current distribution is assumed, which is valid if the harmonics in the air-gap field are of negligible magnitude.

The maximum value of end-ring current is obtained as the product of the average bar current and half the number of bars (or rotor slots) per pole. Assuming that the end-ring current, like the bar current, varies sinusoidally, the root-mean-square value of the end-ring current (I_{er}) becomes:

$$I_{er} = \frac{2}{\pi}\sqrt{2}I_b \times \frac{S_r/2}{p} \times \frac{1}{\sqrt{2}} = \frac{I_b S_r}{\pi p} \qquad (5.28)$$

where I_b is the root-mean-square value of the bar current.

The ventilation is generally better for the end-rings than for the rotor bars, and the current density in the end-ring may be taken equal to or slightly higher than the current density in the bars. However, in most cases the end-ring dimensions are so chosen that its contribution to the rotor winding resistance is minimal.

The cross-section area of each end-ring (a_{er}) is:

$$a_{er} = \frac{I_{er}}{\delta_{er}} = \frac{I_b S_r}{\pi p \delta_{er}} \quad \text{mm}^2 \qquad (5.29)$$

where δ_{er} is the current density in the end-ring, A/mm^2.

The total area of the bars section is:

$$A_{cr} = \frac{I_b S_r}{\delta_b} \qquad (5.30)$$

where δ_b is the current density in the rotor bars, A/mm^2.

Substituting the value I_b from equation (5.30) into equation (5.29) we obtain the cross-section area of each end-ring in terms of the total rotor conductor section

(A_{cr}) as follows:

$$a_{er} = \frac{A_{cr}}{\pi p} \frac{\delta_b}{\delta_{er}}$$

The resistance of the two end-rings is:

$$r_{er} = 2\frac{\rho \pi D_{er}}{a_{er}} \tag{5.31}$$

where D_{er} is the average diameter of the end-ring.

Utilising equations (5.27), (5.28) and (5.31), the total rotor copper loss can be written on the form:

$$I_b^2 r_b + I_{er}^2 r_{er} = \rho I_b^2 S_r^2 \left[\frac{L_b}{S_r a_b} + \frac{2D_{er}}{\pi p^2 a_{er}} \right] \tag{5.32}$$

Dividing equation (5.32) by the bar current squared we get the total rotor resistance as follows:

$$R_r = \rho S_r^2 \left[\frac{L_b}{S_r a_b} + \frac{2D_{er}}{\pi p^2 a_{er}} \right] \quad \Omega$$

The rotor resistance must be expressed in terms of the stator winding before we can utilise it in the motor's equivalent circuit. At standstill, the induction motor acts like a transformer. The rotor resistance referred to the primary voltage (viewed from the stator terminals) is therefore equal to the total rotor resistance times the square of the ratio of the effective rotor turns to the effective stator turns; but what is the effective rotor turns?

The number of phases of the squirrel-cage winding is equal to the number of bars (or rotor slots) per pole, the number of turns in series per phase is equal to the number of pole pairs and the winding factor of the cage winding is equal to 1.0. Therefore, the transformation ratio is:

$$\left[\frac{m_s T_s K_{ws}}{m_r T_r K_{wr}} \right]^2 = \left[\frac{m_s T_s K_{ws}}{(S_r/p)(p/2)} \right]^2 = \left[\frac{2m_s T_s K_{ws}}{S_r} \right]^2$$

The resistance of a squirrel-cage winding per phase in terms of the stator winding is then:

$$r_r' = 4m_s T_s^2 K_{ws}^2 \rho \left[\frac{L_b}{S_r a_b} + \frac{2D_{er}}{\pi p^2 a_{er}} \right]$$

When the radial width of the end-ring is large in comparison to the bar cross-section area, as is the case in small motors, the end-ring component of the cage resistance must be corrected to take into account the non-uniform current distribution in the ring. The end-ring correction factor, K_R, may be taken from

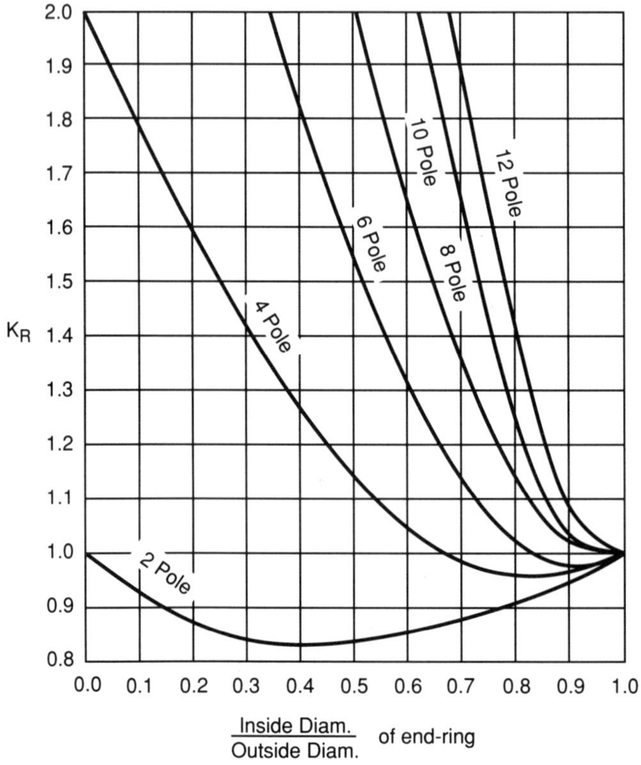

Figure 5.11 End-ring correction factor.

Figure 5.11.

For three-phase motors, the cage winding resistance per phase referred to the stator winding is:

$$r_r' = 12T_s^2K_{ws}^2\rho\left[\frac{L_b}{S_r a_b} + \frac{2D_{er}}{\pi p^2 a_{er}} K_R\right] \tag{5.33}$$

5.3.8 Design of a wound-rotor

The rotor winding for slip-ring motors have the same number of phases and poles as the stator winding. The number of rotor slots must permit a balanced winding. Generally, an integral number of slots per pole per phase are used for the rotor.

Wound-rotor motors are started with the rated voltage applied to the stator winding with external resistances added in the rotor phases. The main field is therefore of normal strength during the starting period. Selection of rotor slot number is not as critical as is the case with squirrel-cage motors. However, to avoid magnetic noises and excessive flux pulsation in the air-gap, the number of rotor slots should be, whenever possible, from 15 to 30% larger or smaller than the number of stator slots.

The number of rotor turns in series per phase determines the voltage between slip-rings, which is at a maximum when the rotor is stationary with the slip-rings open and rated voltage applied to the stator winding. The rotor voltage on open circuit should not exceed 400 V for general purpose small motors. In terms of the stator emf (E_s) and turns in series per phase (T_s), the number of rotor winding turns in series per phase (T_r) is:

$$T_r = \frac{K_{ws}}{K_{wr}} \frac{E_r}{E_s} T_s \tag{5.34}$$

where K_{ws} and K_{wr} are the stator and rotor winding factors, respectively, T_s is the stator turns per phase and E_r is the rotor open circuit voltage per phase, which should not exceed 400 V or 230 V for delta connected and star-connected rotors; respectively.

The cross-section area of the rotor conductor can be found by assuming a suitable value of current density in the rotor winding. In order to avoid excessive rotor copper loss and heating of the insulated winding, the current density in the rotor should be assumed equal to the current density of the stator winding. The required slot area can be estimated by assuming a slot space factor similar to that of the stator winding (about 0.4), to allow for space required for slot insulation and wedges.

The rotor windings of small slip-ring motors are usually of the single layer mush type. It is common practice to use several wires in parallel to obtain the required conductor area. This facilitates placing the winding in the partially closed slots. The rotor winding is usually star connected and its three leads are brought through the shaft to the slip-rings.

The rotor winding resistance of a slip-ring motor per phase is calculated in a similar manner as the stator winding resistance as follows:

$$r_r = \frac{\rho T_r L_{mtr}}{a_r \times 10^{-6}}$$

where a_r is conductor cross- section area, mm^2 and ρ is the resistivity, Ωm.

The rotor resistance per phase referred to stator is:

$$r_r' = \left(\frac{K_{ws} T_s}{K_{wr} T_r}\right)^2 \frac{\rho T_r L_{mtr}}{a_r \times 10^{-6}} \tag{5.35}$$

5.3.9 Rotor tooth and core

For constant speed induction motors, the frequency of the flux reversal in the rotor is very low. The core losses will be small, almost negligible, even if the flux density is high. However, the maximum flux density in the teeth can only be little higher than that used for the stator teeth (1.7 T). It is limited by the ampere-turns required to pass the flux through the teeth.

The flux density in the rotor core is generally equal to the stator core density and it should not exceed 1.5 T. As the rotor core carries half the pole flux, the minimum core depth can be obtained with the aid of equation (5.17).

5.4 OPERATING CHARACTERISTICS

5.4.1 No load current

The no load current of an induction motor is made up of two components: the magnetising current, I_m, which is 90° out of phase with respect to the applied voltage and the loss component, I_p, which is in phase with the voltage.

The magnetising current is obtained from equation (A3.2), Appendix 3, by calculating the total ampere-turns per pole required to pass the flux in the magnetic circuit. The calculation of the ampere-turns per pole is described in Appendix 3.

The calculation of the loss component of the no load current involves the determination of the no load losses. These are the iron losses in the teeth and cores and the friction and windage losses.

The iron losses in the stator teeth and core are determined by calculating their respective weight and multiplying it with the specific loss (per unit weight) corresponding to the flux density values. The specific loss curves of the laminations are given in Appendix 1.

As the rotor frequency is very low, the rotor iron losses may be ignored. However, in some wound-rotor designs when the motor is expected to run at high slips during its loading cycle, the rotor iron losses should be included in the loss calculations.

The friction and windage losses depend on the motor's construction and accurate values are very difficult to estimate in the general case. In the absence of test data on similar machines, the friction and windage losses may be taken between 1 and 3% of the rated output.

The loss component of the no load current per phase for three-phase motors is:

$$I_p = \frac{\text{Total no-load losses}}{3E_s} \tag{5.36}$$

and the no load current per phase is:

$$I_0 = \sqrt{I_m^2 + I_p^2} \tag{5.37}$$

Usual values of the no load current for small induction motors lies between 30 and 50% of the full load current.

5.4.2 Equivalent circuit parameters

The calculations of stator and rotor resistances are described above. The leakage reactance calculation is described in Appendix 3.

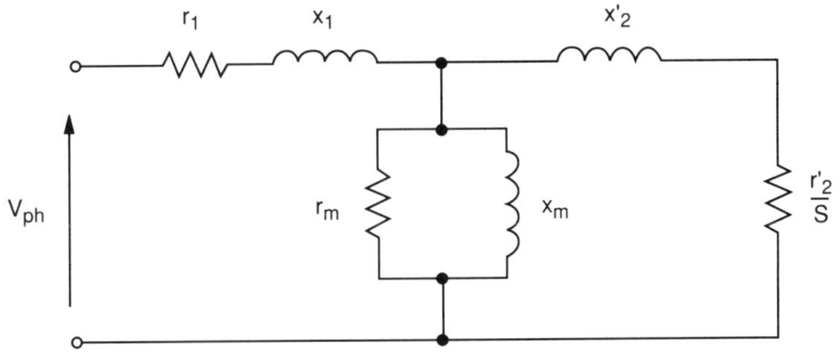

Figure 5.12 Induction motor equivalent circuit.

The magnetising reactance per phase, x_m, can be calculated when the magnetising current (I_m) and stator leakage reactance (x_1) become known, as follows:

$$x_m = \frac{V_{ph}}{I_m} - x_1 \tag{5.38}$$

The equivalent circuit method may be used to determine the induction motor performance characteristics. Figure 5.12 shows the equivalent circuit of a single-cage induction motor. This concept, however, does not take magnetic saturation into account. As induction motors become heavily saturated during starting, the equivalent circuit method may only be used to predict performance characteristics at slips not greater than that corresponding to the pull-out torque of a normal design. A method in which the conventional equivalent circuit of an induction motor is utilised in conjunction with a finite element non-linear model to predict the motor's performance is described in chapter 7 [5.8].

5.5 SIZES OF INDUCTION MOTOR LAMINATIONS

A wide range of stator and rotor laminations is available and, practically, all standard designs can be accommodated. An example of sizes of induction motor laminations is given in Table 5.2 [5.1]. In the following section, sample designs of induction motors are given and, whenever possible, laminations from the range given in Table 5.2 are utilised.

5.6 SAMPLE DESIGNS

5.6.1 Design of a 50 Hz motor

Design of a 0.25 kW, 1380 rev/min, 415 V, star connected, 50 Hz, three-phase, 4-pole, squirrel-cage induction motor. The machine is meant to be sold in a highly competitive market and it is to be insulated to class F insulation.

Table 5.2 Small induction motor laminations [5.1]

O/D (mm)	I/D (mm)	Slots (S_s/S_r)	O/D (mm)	I/D (mm)	Slots (S_s/S_r)
35.00	17.50	12/11	90.00	50.00	24/17
36.50	16.00	12/10	90.00	50.00	12/17
39.20	20.00	12/10	90.00	50.00	24/33
41.60	22.00	18/17	90.00	55.00	36/30
42.00	21.00	12/19	100.00	50.00	18/22
42.00	21.70	12/11	100.00	50.00	12/15
48.20	22.00	12/10	100.00	50.00	24/22
48.25	22.05	10/12	99.78	55.00	24/17
50.00	27.00	12/16	100.00	55.00	24/17
52.20	27.00	12/16	106.50	55.00	24/18
55.00	28.00	16/18	106.50	55.00	24/17
56.00	30.00	12/14	106.50	62.00	24/22
56.30	28.00	12/15	106.50	70.00	36/30
58.00	25.00	12/16	110.00	55.00	24/18
58.25	27.02	12/16	110.00	60.00	24/22
60.00	30.00	12/15	110.00	60.00	36/30
60.00	32.00	12/15	110.00	60.00	24/18
65.00	35.00	12/16	120.00	62.00	24/18
65.00	37.50	24/18	120.00	62.00	24/17
65.10	35.00	12/16	120.00	70.00	24/22
65.20	30.00	12/17	120.00	70.00	24/30
69.00	35.00	12/16	120.00	70.00	36/28
69.00	37.00	12/17	120.00	80.00	36/30
70.00	34.00	12/13	120.00	80.00	36/26
70.00	35.00	12/17	125.00	60.00	24/18
70.10	37.00	12/17	130.00	80.00	36/28
73.00	40.00	16/17	135.00	70.00	24/18
73.10	40.07	16/20	135.00	70.00	24/17
75.00	40.00	12/17	135.00	80.00	24/30
80.00	39.95	12/17	135.00	80.00	36/28
80.00	40.00	12/14	135.00	90.00	36/26
80.00	40.00	18/22	135.00	90.00	36/30
80.00	40.00	16/22	135.00	80.00	36/30
80.00	40.00	12/14	135.00	90.00	36/44
80.00	40.00	12/18	145.00	85.00	36/28
80.00	45.00	24/22	150.00	80.00	24/18
80.00	50.00	36/30	150.00	80.00	24/17
80.10	45.00	16/22	150.00	90.00	24/30
80.10	45.00	12/22	150.00	90.00	36/28
85.00	45.50	18/23	150.00	90.00	36/44
90.00	45.00	18/22	150.00	103.00	36/34
90.00	45.00	12/18	160.00	100.00	40/44
90.00	45.00	24/18	165.00	80.00	24/30
90.00	45.00	24/17	165.00	103.00	24/30
90.00	50.00	24/22	165.00	103.00	36/30
90.00	50.00	24/22	165.00	103.00	36/44

Continued

Table 5.2 *(Continued)*

O/D (mm)	*I/D (mm)*	*Slots (S_s/S_r)*	O/D (mm)	I/D (mm)	Slots (S_s/S_r)
165.00	103.00	48/44	230.30	150.00	36/44
165.00	110.00	36/44	240.00	135.00	36/30
170.00	90.00	24/18	240.00	150.00	36/28
170.00	90.00	36/30	240.00	150.00	36/32
170.00	103.00	36/28	240.00	165.00	54/38
170.00	103.00	36/30	240.00	170.00	48/44
170.00	115.00	36/34	270.00	150.00	36/40
180.00	110.00	36/44	270.00	170.00	36/28
195.00	125.00	36/30	270.00	170.00	36/32
200.00	110.00	36/30	270.00	180.00	54/38
200.00	125.00	36/28	270.00	190.00	48/44
200.00	125.00	36/32	300.00	170.00	36/40
200.00	135.00	36/34	300.00	190.00	48/40
200.00	140.00	48/44	300.00	200.00	54/38
220.00	130.00	36/44	340.00	190.00	36/40

Main dimensions

For this size, the acceptable values of efficiency and power factor are about 0.7 each, therefore:

$$Q = \frac{0.25}{0.7 \times 0.7} = 0.51\,\text{kVA}$$

Let:

$$B_{av} = 0.5\,\text{T} \quad \text{and} \quad ac = 22{,}000 \text{ ampere-conductor/m}$$

Initial value of $K_{w\varepsilon}$ is assumed 0.955, the output coefficient is:

$$C_o = 1.11\pi^2 \times 0.5 \times 22{,}000 \times 0.955 \times 10^{-3} = 115$$

The synchronous speed is 25 rev/s. The product D^2L is:

$$D^2L = \frac{0.51}{115 \times 25} = 1.77 \times 10^{-4}\,\text{m}^3$$

For a low cost design the core length to pole pitch ratio should be between 1.5 and 2.0. Assuming L/τ of 1.5 yields a diameter and core length of 53 mm and 63 mm, respectively.

In order to keep the cost low, we must utilise standard laminations for this design. We use K&S lamination with 55 mm diameter. Therefore:

$$D = 55\,\text{mm}, \quad \text{and} \quad L = 58\,\text{mm}$$

and:

$$\frac{L}{\tau} = 1.34$$

The lamination thickness is 0.5 mm and the staking factor is 0.95. The nett iron length is 55.1 mm. The pole pitch is 43.2 mm. The flux per pole is:

$$\phi = 0.5 \times 0.0551 \times 0.043 = 1.18\,\text{mWb}$$

The air-gap length is determined with the aid of equation (5.7):

$$l_g = 3.06 - \frac{6560}{55 + 2280} = 0.25\,\text{mm}$$

Stator design

The number of slots per pole per phase (q) should be an integer not less than 2. With q = 2, the number of stator slots will be 24 and the slot pitch on the air-gap diameter will be 7.2 mm. This is quite small and if we choose a larger number of slots it will be even smaller. Therefore, we use 2 slots per pole per phase. The corresponding value of winding distribution factor is 0.966. Semiclosed tapered slots are used.

The emf induced in the stator winding may be taken equal to 0.97 times the phase value of the applied voltage. Therefore:

$$E_s = 0.97 \times \frac{415}{\sqrt{3}} = 232.4\,\text{V}$$

Equation (5.10) is utilised to calculate the number of turns in series per phase as follows:

$$T_s = \frac{232.4}{4.44 \times 50 \times 1.18 \times 10^{-3} \times 0.966} = 918$$

The total number of stator conductors is 5508 and the number of conductors per slot is 229.5. Therefore, we use 230 conductors per slot. The number of turns in series per phase will be 920. Mush winding is used for this motor and the number of coils is equal to half the number of stator slots, or 12.

The coil span must be an odd number for the mush winding. The pole pitch is 6 slots and the slot angle is 30 electrical degrees. Therefore, we use a coil pitch of 5 slots, or 150 electrical degrees. The winding is short-pitched by one slot and the pitching factor is:

$$K_c = \sin\frac{150}{2} = 0.966$$

The stator winding factor is now known:

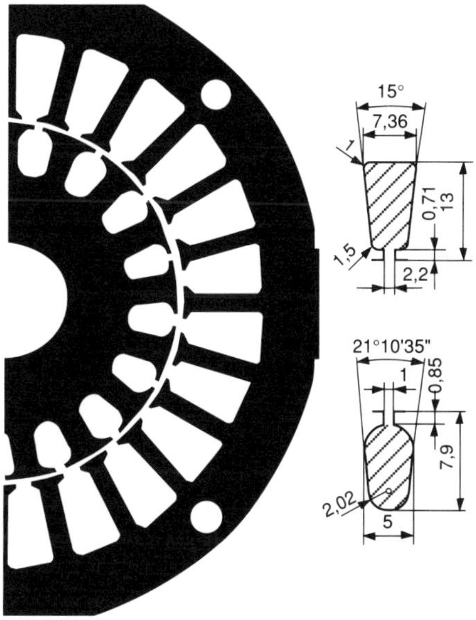

Figure 5.13 Stator and rotor laminations.

$$K_{ws} = K_{ds}K_{cs} = 0.966 \times 0.966 = 0.933$$

The value of the flux per pole and air-gap density should be corrected to:

$$\phi = \frac{232.4}{4.44 \times 50 \times 920 \times 0.933} = 1.22\,\text{mWb} \quad \text{and} \quad B_{av} = \frac{\phi}{\tau L_i} = 0.51\,\text{T}$$

The stator phase current is:

$$I_s = \frac{0.51 \times 10^3}{3 \times 240} = 0.7\,\text{A}$$

The machine is to be insulated with class F insulation and, therefore, we can utilise high current densities. Let the stator winding current density be 6 A/mm^2. The conductor cross-section area is:

$$a_s = \frac{0.7}{6.0} = 0.118\,\text{mm}^2$$

The required conductor diameter is 0.39 mm. The nearest standard wire has a diameter of 0.4 mm and its cross-section area is 0.1257 mm^2.

One of the K&S laminations with 55 mm diameter is shown in Figure 5.13 [5.1]. The advantage of using these laminations lies in the slot combination. If we can use these laminations no skewing will be required.

The total area of 230 conductors per slot is about 28.9 mm^2. The slot area of the K&S lamination shown in Figure 5.13 is 72.6 mm^2. The slot space factor is about 0.4.

Table 5.3 Details of stator and rotor laminations

Part	O/D (mm)	I/D (mm)	A_s (mm^2)	W_t (mm)
Stator	99.78	55.00	72.60	3.00
Rotor	54.50	18.00	27.70	3.90

The length of the mean turn of the stator winding is calculated with the aid of equation (5.14) as follows:

$$L_{mts} = 2 \times 0.058 + 2.3 \times 0.43 = 0.215\,\text{m}$$

The stator resistance per phase is obtained from equation (5.16). At 25 °C it is:

$$r_s = \frac{1.8 \times 10^{-8} \times 920 \times 0.215}{0.1257 \times 10^{-6}} = 28.4\,\Omega$$

The lamination will be satisfactory if the flux densities in the teeth and cores are satisfactory. The width of the stator tooth of the K&S lamination shown in Figure 5.13 is 3 mm. The tooth density is:

$$B_{ts} = \frac{1.22 \times 10^{-3}/6}{0.0551 \times 0.003} = 1.23\,\text{T}$$

This is satisfactory as the tooth is neither saturated nor being worked at a low flux density.

The stator core depth is 9.39 mm and the core flux density is:

$$B_{cs} = \frac{1.22 \times 10^{-3}/2}{0.0551 \times 9.39 \times 10^{-3}} = 1.28\,\text{T}$$

which is quite acceptable.

Rotor design

The rotor diameter is:

$$D_r = D - 2l_g = 55 - 2 \times 0.25 = 54.5\,\text{mm}$$

If we take the number of rotor slots 30% less than the number of stator slots we need 16.8 slots. Therefore 17 rotor slots are quite acceptable on this account. The number of stator slots of 24 and rotor slots of 17 satisfies the criteria discussed in section 5.3.6 above. This combination is available from K&S and the stator and rotor lamination details are given in Table 5.3.

Therefore:

$$S_s = 24 \quad \text{and} \quad S_r = 17$$

The area of the rotor bar is 27.7 mm^2 and the area of the 17 bars is about 471 mm^2.

We can establish whether the bar area is satisfactory or not by finding the ratio of the total rotor conductor area to the stator copper area; a value between 0.5 and 0.8 is likely to produce a satisfactory design.

The total stator copper area (the cross-section area of 5520 conductors) is about 693 mm^2. Therefore:

$$\frac{A_{cr}}{A_{cs}} = \frac{471}{693} = 0.68$$

which is acceptable.

Next, we examine the current density in the rotor bars. The rotor bar current is:

$$I_b = \frac{2m_s K_{ws} T_s}{S_r} \times I_s \cos\phi = \frac{2 \times 3 \times 0.933 \times 920}{17} \times 0.73 \times 0.7 = 154.9\,\text{A}$$

and the current density is about 5.6 A/mm^2 which is quite satisfactory.

Before finally accepting this rotor lamination, the rotor tooth and core flux density values need to be examined. The tooth width is 3.9 mm and the tooth flux density is:

$$B_{tr} = \frac{1.22 \times 10^{-3}}{17/4 \times 0.0551 \times 3.9 \times 10^{-3}} = 1.34\,\text{T}$$

The depth of the rotor core is 11.8 mm and the core density is about 0.9 T. Therefore, both the rotor tooth and core flux density are acceptable.

The end-ring current is calculated from equation (5.28) as follows:

$$I_{er} = \frac{154.9 \times 17}{\pi \times 4} = 209.5\,\text{A}$$

Taking the current density in the end-ring as 5.6 A/mm^2 (equal to the bar current density), the cross-section area of each end-ring becomes 37.4 mm^2. The outer diameter of the end-ring must be smaller than the rotor diameter. Let the end-ring diameter be 53 mm, its radial thickness be 8 mm and its axial length be 10 mm. The end-ring area will be 80 mm^2, about twice that required to achieve a current density value of 5.6 A/mm^2. These dimensions would be more suitable for manufacturing. The current density in the end-ring becomes only about 2.6 A/mm^2. The rotor will be die-cast using aluminium.

The mean diameter of the end-ring is 49 mm. The end-ring factor is obtained from Figure 5.11 and it is 0.95. The cage winding resistance per phase referred to stator is calculated with the aid of equation (5.33) as follows:

$$r' = 12 \times 920^2 \times 0.933^2 \times 2.7 \times 10^{-2}\left[\frac{0.058}{17 \times 27.7} + \frac{2 \times 0.049 \times 0.95}{\pi \times 16 \times 80}\right]$$

$$= 34.9\,\Omega$$

Table 5.4 Magnetisation data

Section	B (T)	B_{60} (T)	Length (mm)	at/m	AT
Air-gap	0.51	0.69	0.349	55.2×10^4	192.6
Stator teeth	1.23	1.67	13.00	280	3.6
Stator core	1.28	—	23.6	250	5.9
Rotor teeth	1.34	1.82	7.9	1000	7.9
Rotor core	0.9	—	7.0	160	1.1

Table 5.5 Iron losses in teeth and core

Section	Weight (kg)	B_{max} (T)	p_1 (W/kg)	Iron loss (W)
Stator teeth	0.4	1.93	6.0	2.4
Stator core	1.12	2.0	7,2	8.0

No load current

The method of calculating the ampere-turns per pole required to pass the flux in the magnetic circuit is given in Appendix 3. The magnetisation data for the present design is given in Table 5.4.

The magnetising current is:

$$I_m = \frac{0.427 \times 4 \times 211.1}{920 \times 0.933} = 0.42\,\text{A}$$

The calculation of the iron losses in the stator teeth and core are summarised in Table 5.5. The total iron losses are 10.4 W.

The friction and windage losses are taken as 2.5% of the rated output, or 5.75 W. The total no load losses are 16.15 W. The loss component of the no load current is:

$$I_p = \frac{16.15}{3 \times 232.4} = 0.023\,\text{A}$$

and the no load current (I_0) is approximately 0.42 A.

The resistance r_m in the equivalent circuit which represents the iron, friction and windage losses is:

$$r_m = \frac{232.4}{0.23} = 10\,104\,\Omega$$

Leakage reactance

The permeance coefficients for the stator and rotor slots are estimated with the aid of Figure A3.3 (the slot dimensions are given in Figure 5.13):

$$\lambda_{ss} = 1.25\,\mu_0 \quad \text{and} \quad \lambda_{sr} = 1.85\,\mu_0$$

and the leakage reactance due to the stator slots is:

$$x_{ss} = \frac{8 \times \pi \times 50 \times 920^2 \times 0.058}{4 \times 2} \times 1.25\,\mu_0 = 12.1\,\Omega$$

The rotor slot permeance coefficient referred to stator is:

$$\lambda'_{sr} = 1.85\,\mu_0 \frac{0.933^2 \times 24}{1^2 \times 17} = 2.27\,\mu_0$$

and the rotor slot leakage reactance referred to stator is:

$$x'_{sr} = \frac{8 \times \pi \times 50 \times 920^2 \times 0.058}{4 \times 2} \times 2.27\,\mu_0 = 22\,\Omega$$

The overhang factor for the coil span to pole pitch ratio of 5/6 is obtained from Figure A3.4, and it is equal to 0.83. Equation (A3.14) is utilised to calculate:

$$L_0\lambda_0 = \mu_0 \frac{0.83 \times (43.2 \times 10^{-3})^2}{\pi \times 10.07 \times 10^{-3}} = 0.049\,\mu_0$$

The overhang leakage reactance is calculated with the aid of equation (A3.15) as follows:

$$x_0 = \frac{8 \times \pi \times 50 \times 920^2}{4 \times 2} \times 0.049\,\mu_0 = 8.18\,\Omega$$

The zigzag leakage reactance may be calculated from equation (A3.13). The total zigzag leakage per phase is about 1.67 Ω divided almost equally between stator and rotor.

The belt leakage reactance may be calculated with the aid of equation (A3.16). However, this value is negligible.

The total stator and rotor leakage reactance per phase are:

$$x_1 = 21.1\,\Omega \quad \text{and} \quad x'_2 = 22.8\,\Omega$$

The magnetising reactance is:

$$x_m = \frac{239.6}{0.42} - 21.1 = 549.4\,\Omega$$

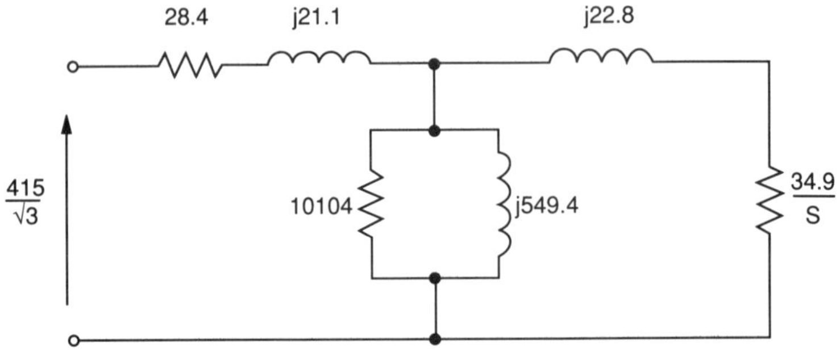

Figure 5.14 Equivalent circuit of a 250 W, 1380 rev/min induction motor.

The values of all the equivalent circuit components are now known and the motor's equivalent circuit is shown in Figure 5.14.

The equivalent circuit may be used to determine the motor's operating characteristics. However, it must be remembered that the equivalent circuit parameters were obtained without taking the effect of magnetic saturation into account. This approximation will affect the accuracy of the predicted starting and run-up performance more than the running characteristics. However, the output power at a slip of 8% (or speed of 1380 rev/min) will be about 325 W. After allowing for saturation and stray losses the output would probably be near 250 W.

5.6.2 Design of a 400 Hz motor

Design of a 0.7 kW, 200 V, 400 Hz, 11,400 rev/min, three-phase induction motor to drive a fuel pump on an aircraft. The efficiency and power factor to be not less than 0.7 and 0.8, respectively.

Main dimensions

The motor is designed to run immersed in aircraft fuel and, therefore, high values of specific loadings can be used from the heat dissipation viewpoint. However, in order to limit the magnetising current to an acceptable value and obtain high power factor we should use moderate values of specific magnetic loading.

Let:

$$B_{av} = 0.4\,\text{T} \quad \text{and} \quad ac = 40{,}000 \text{ ampere-conductor/m}$$

Initial value of K_w may be assumed 0.955 and the output coefficient is:

$$C_o = 1.11\pi^2 \times 0.4 \times 40{,}000 \times 0.955 \times 10^{-3} = 167.4$$

The kVA input is:

$$Q = \frac{0.7}{0.7 \times 0.8} = 1.25\,\text{kVA}$$

The motor has 4 poles and the synchronous speed (n_s) is 200 rev/s. The D^2L product is:

$$D^2L = \frac{1.25}{167.4 \times 200} = 3.734 \times 10^{-5}\,\text{m}^3$$

For good power factor the ratio of the core length to pole pitch should be between 1 and 1.5. For 4-pole motors if we take $L = D$, the L/τ ratio will be 1.27. Let:

$$D = 33\,\text{mm} \quad \text{and} \quad L = 34\,\text{mm}$$

Stator design

The cost is not the most important design criterion in this design and we utilise low loss lamination material. The lamination thickness will be 0.35 mm or 0.27 mm and the staking factor is typically 0.95. The nett iron length is 32.3 mm and the pole pitch is 26 mm. Therefore, the flux per pole is 3.348×10^4 Wb. The induced emf per phase is taken as 0.97 times the phase terminal voltage or 112 V. The number of stator turns per phase is:

$$T_s = \frac{112}{4.44 \times 400 \times 3.348 \times 10^{-4} \times 0.955} = 197$$

Because of the small diameter, we choose 2 slots per pole per phase. The number of stator slots is 24 and the distribution factor K_d is 0.966. Single layer mush winding is used and the coil span is 5 slots, or 150 electrical degrees. The pitching factor is 0.966 and the stator winding factor is 0.933.

Now, the number of stator turns must be corrected to give an integer number of conductors per slot. With 50 conductors per slot the required turns per phase will be 200. The value of flux per pole becomes 3.38×10^{-4} Wb and the average air-gap density (B_{av}) remains approximately unchanged at 0.4 T.

The stator phase current is:

$$I_s = \frac{0.7 \times 10^3}{3 \times 115.4 \times 0.7 \times 0.8} = 3.6\,\text{A}$$

We can, indeed, we must, design for a much higher current density than used for air-cooled motors. Assuming a stator current density of 20 A/mm^2, the conductor cross-section area will be:

$$a_s = \frac{3.6}{20} = 0.18\,\text{mm}^2$$

The required conductor diameter is 0.48 mm. Therefore, we use a standard copper wire with 0.5 mm diameter and its cross-section area is 0.196 mm^2. The stator current density is now about 18.4 A/mm^2.

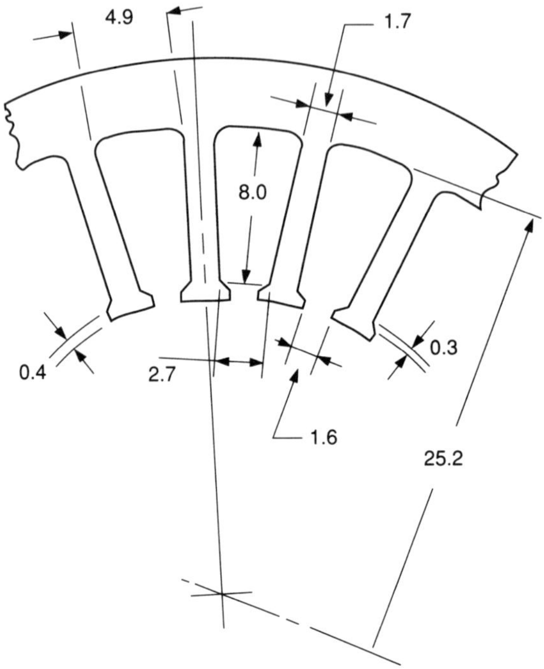

Figure 5.15 Details of stator lamination.

The length of the mean stator turn is:

$$L_{mts} = 2 \times 34 + 2.3 \times 26 = 127.8\,\text{mm}$$

The stator resistance per phase is:

$$r_s = \frac{1.8 \times 10^{-8} \times 200 \times 0.1278}{0.196 \times 10^{-6}} = 2.35\Omega$$

The total area of 50 conductors is about 9.8 mm^2 and a slot area of between 25 and 32 mm^2 yields a slot space factor of between 0.4 and 0.3. The stator lamination is shown in Figure 5.15 and the slot area is 30.4 mm^2. This lamination will be acceptable if the tooth flux density is not too high. The stator teeth flux density is:

$$B_{ts} = \frac{3.38 \times 10^{-4}}{6 \times 1.7 \times 10^{-3} \times 32.3 \times 10^{-3}} = 1.03\,\text{T}$$

which is acceptable.

The depth of stator core is chosen so that the core density is about 1.5 T, therefore:

$$d_{cs} = \frac{3.38 \times 10^{-4}}{2 \times 1.5 \times 32.3 \times 10^{-3}} = 3.5 \times 10^{-3}\,\text{m} \quad \text{or} \quad 3.5\,\text{mm}$$

The stator outer diameter is 57.4 mm.

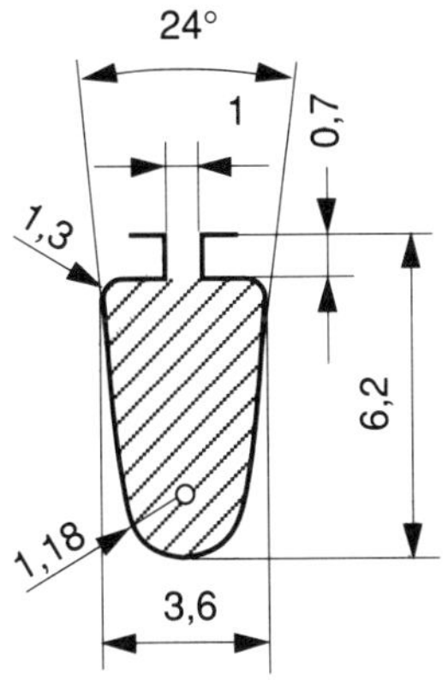

Figure 5.16 Rotor slot dimensions.

Rotor design:

The air-gap radial length is:

$$l_g = 3.06 - \frac{6560}{33 + 2280} = 0.22\,\text{mm}$$

and the rotor diameter is 32.56 mm.

The number of rotor slots should be between 15 and 30% greater or less than the number stator slots. Because of the small rotor diameter we choose rotor slots less than stator slots. Choosing 15 rotor slots satisfies the criterion discussed above relating to slot harmonics. The rotor slot is shown in Figure 5.16 and the tooth width is 2.2 mm.

The rotor bar current is:

$$I_b = \frac{2m_s K_{ws} T_s}{S_r} I_s \cos\phi = \frac{2 \times 3 \times 0.933 \times 200}{15} = 215\,\text{A}$$

and the current density is about 14.2 A/mm^2.

The end-ring current is:

$$I_{er} = \frac{215 \times 15}{4\pi} = 267\,\text{A}$$

Although the cost of this particular motor is not the most important criterion, if the rated power can be obtained with a diecast aluminium cage winding, its manufacturing cost will be considerably reduced.

The flux density in the rotor teeth is:

$$B_{tr} = \frac{3.38 \times 10^{-4}}{15/4 \times 32.3 \times 10^{-3} \times 2.2 \times 10^{-3}} = 1.27\,\text{T}$$

The shaft diameter is 10 mm and the depth of the rotor core is 5.08 mm. The flux density in the rotor core is:

Table 5.6 Magnetisation data

Section	B (T)	B_{60} (T)	Length (mm)	at/m	AT
Air-gap	0.40	0.544	0.300	43.29×10^4	129.9
Stator teeth	1.03	1.40	8.70	23	0.2
Stator core	1.50	—	14.1	32	0.45
Rotor teeth	1.27	1.73	6.2	160	1.0
Rotor core	1.03	—	3.95	11	0.05

Table 5.7 Iron losses in teeth and core

Section	Weight (kg)	B_{max} (T)	p_1 (W/kg)	Iron loss (W)
Stator teeth	0.088	1.620	37	3.25
Stator core	0.150	2.350	86.5	12.9

$$B_{cr} = \frac{3.38 \times 10^{-4}}{2 \times 32.3 \times 10^{-3} \times 5.08 \times 10^{-3}} = 1.03\,\text{T}$$

The flux densities in the teeth and core are acceptable and we will aim to utilise a diecast cage winding.

Let the end-ring outer and inner diameters be 31.2 mm and 12 mm, respectively. The end-ring factor corresponding to these values is 1.28. Let the axial dimension of the end ring be 10 mm. Therefore, the end-ring area is 96 mm^2.

The rotor resistance referred to stator is:

$$r'_r = 12 \times 200^2 \times 0.933^2 \times 2.7 \times 10^{-2}\left[\frac{0.034}{15 \times 15.1} + \frac{2 \times 0.0216 \times 1.28}{\pi \times 16 \times 96}\right] = 1.82\,\Omega$$

No load current

The magnetisation data is summarised in Table 5.6 and the magnetising current is:

$$I_m = \frac{0.427 \times 4 \times 131.6}{200 \times 0.933} = 1.2\,\text{A} \quad \text{or } 33\% \text{ of } I_{fl}$$

The iron losses in the stator teeth and core are given in Table 5.7. The drag loss as a percentage of the rated output is much higher than the windage loss in dry motors and it should be determined experimentally from tests on similar motors. In the absence of typical drag characteristics the friction and drag losses may be taken between 4% and 8% of the rated output.

Taking the friction and drag losses as 5% of the rated output or 35 W, the loss component of the no load current is:

$$I_p = \frac{16.15 + 35}{3 \times 112} = 0.15\,\text{A}$$

and the no load current is approximately 1.2 A.

The resistance r_m representing the iron, friction and drag losses is:

$$r_m = \frac{112}{0.15} = 746\,\Omega$$

Leakage reactance

The permeance coefficients of the stator and rotor slots are:

$$\lambda_{ss} = 1.25\,\mu_0 \quad \text{and} \quad \lambda_{sr} = 0.75\,\mu_0$$

The stator slot leakage reactance is:

$$x_{ss} = \frac{8 \times \pi \times 400 \times 200^2 \times 0.034}{4 \times 2} \times 1.25\,\mu_0 = 2.68\,\Omega$$

The overhang factor is 0.83 and:

$$L_o\lambda_0 = \mu_0 \frac{0.83 \times 0.026^2}{\pi \times 4.32 \times 10^{-3}} = 0.04\,\mu_0$$

The overhang leakage reactance is:

$$x_o = \frac{8 \times \pi \times 400 \times 200^2}{4 \times 2} \times 0.04\,\mu_0 = 2.6\,\Omega$$

Equation (A3.13) is used to calculate the zigzag leakage reactanceat 0.135 Ω and the belt leakage reactance is negligible. Therefore:

$$x_1 = 5.35\,\Omega$$

The rotor slot permeance coefficient referred to stator is:

$$\lambda'_{sr} = 0.75\,\mu_0 \frac{0.933^2 \times 24}{15} = 1.04\,\mu_0$$

and the rotor slot leakage reactance referred to stator is 2.23 Ω. Assuming no skewing is necessary, the rotor leakage reactance is:

$$x'_2 = 2.3\,\Omega$$

The magnetising reactance is:

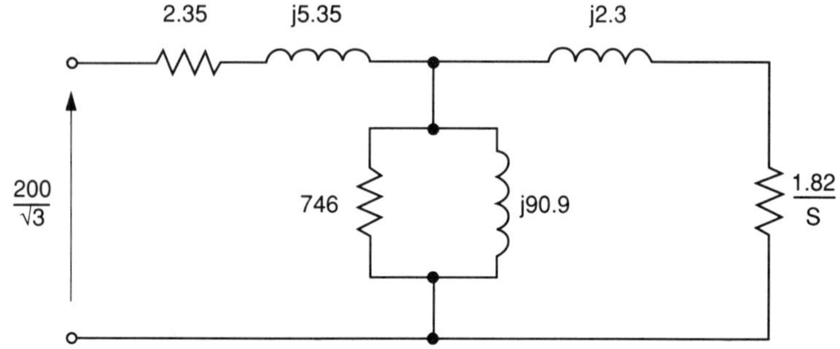

Figure 5.17 Equivalent circuit of a 700 W, 11,400 rev/min induction motor.

$$x_m = \frac{115.5}{1.2} - 5.35 = 90.0\,\Omega$$

The motor's equivalent circuit is shown in Figure 5.17. When the equivalent circuit is used to predict the performance characteristics of the motor, the output power at 5% slip (or a speed of 11,400 rev/min) is found to be about 790 W, some 90 watts greater than the required output. Bearing in mind the approximate nature of the calculations and that we assumed the drag losses to be 5% of the rated output, this design would probably produce the required output at the rated speed.

5.7 REFERENCES

[5.1] Kienle and Spiess (1985) *Stator and Rotor Laminations for Induction Motors*, Stanz- und Druckgie*ß*werk GmbH, D-7123 Sachsenheim 1, Germany.
[5.2] McPherson, G. and Laramore, R. D. (1990) *An Introduction to Electrical Machines and Transformers*, New York: John Wiley & Sons.
[5.3] Finzi, L. A. and Paice, D. A. (1968) *IEEE Trans.*, **PAS-87**, No. 2, p. 585.
[5.4] Chalmers, B. J. and Herman, J. (1977) *Proc. IEE, Pt. B*, **124**, No. 5, p. 449.
[5.5] Sambath, H. P. (1976) *PhD Thesis*, University of Manchester Institute of Science and Technology, Manchester, England.
[5.6] Chalmers, B. J. and Hamdi, E. S. (1981) *IEE Conference Publication No. 202*, The Institution of Electrical Engineers, London, England, p. 16.
[5.7] Chalmers, B. J. and Hamdi, E. S. (1981) *Proc. IEE, Pt. B*, **129**, No. 1, p. 27.
[5.8] Belmans, R. *et al* (1990) *IEEE Trans. on Energy Conversion*, **5**, No. 4, p. 719.

6

PERMANENT MAGNET MOTORS

There are many permanent magnet materials which have been successfully utilised in electromechanical devices in general, and in motors in particular. The generic magnet groups and their characteristics are discussed in chapter 1 of this book.

Alnico has the advantages of higher flux density and lower temperature dependence over ceramic magnets. On the other hand, Alnico magnets have lower resistance to demagnetisation and higher cost. Rare earth magnets offer outstanding magnetic properties but they are, generally, more expensive than Alnicos and ceramics. The application requirements dictate which magnet material to be used. As well as the required magnetic performance, when selecting a magnet material the designer must also consider operating temperature range, external demagnetising fields, weight, cost and space limitations.

Motors of general commercial characteristics utilise ceramic or polymer-bonded neodymium–iron–boron magnets while sintered rare earth magnets are reserved for high performance motors and for applications where the size and weight are restricted. Alnicos are used for some applications when the motor is required to operate in environmental extremes.

6.1 MAGNET DESIGN

At present, computer-aided design methods involving numerical field solutions are widely used to optimise the design of permanent magnet devices. However, the dimensions of a permanent magnet employed in an electromechanical device to furnish a specified value of air-gap flux density may be estimated with the aid of a few simple equations. Quick and simple hand calculations are useful in assessing the feasibility of a design, or in providing a good starting point for an iterative CAD procedure.

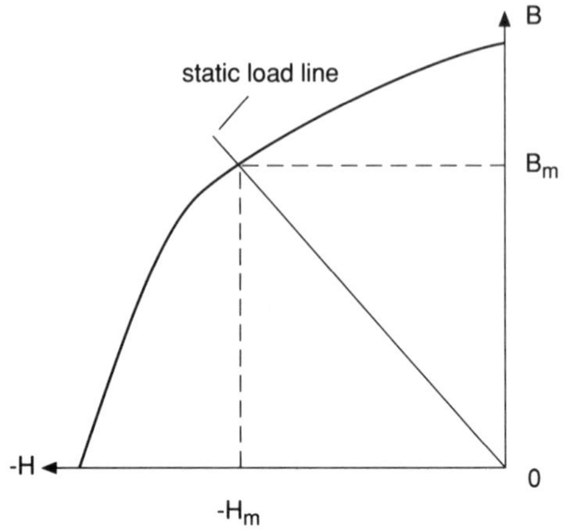

Figure 6.1 Demagnetisation curve and static load line.

6.1.1 Magnet dimensions

There are two basic equations from which the dimensions of a permanent magnet can be estimated. The first is the mathematical representation of the assumption that all the magnetic flux lines produced by the magnet, less the leakage flux, cross the air-gap; therefore:

$$B_m A_m = K_l B_g A_g \tag{6.1}$$

where B_m and A_m, B_g and A_g are the flux density and area of the magnet and air-gap, respectively, and K_l is the leakage factor. The magnet's magnetomotive force ($H_m l_m$) passes the flux in the magnetic circuit, and most of it is required to pass the magnetic flux lines across the air-gap, therefore:

$$H_m l_m = K_r H_g l_g \tag{6.2}$$

where H_g and l_g are the air-gap magnetic field intensity and length. K_r is called the reluctance factor, and it accounts for the magnetomotive force required to pass the flux lines in the soft iron parts of the magnetic circuit.

If one assumes an operating point B_m at, say, 0.5 B_r, the magnet surface area can be readily determined from equation (6.1). H_m corresponding to the assumed B_m can be determined from the demagnetisation curve of the magnet material, as illustrated in Figure 6.1, and the magnet length (l_m) can be calculated from equation (6.2). Alternatively, if the magnet and air-gap dimensions are known, the magnet operating point on the demagnetisation curve can be determined by rearranging equations (6.1) and (6.2) to obtain expressions for B_m and H_m and dividing to obtain:

$$\frac{B_m}{H_m} = \mu_0 \frac{K_l}{K_r} \frac{A_g}{A_m} \frac{l_m}{l_g} \tag{6.3}$$

or,

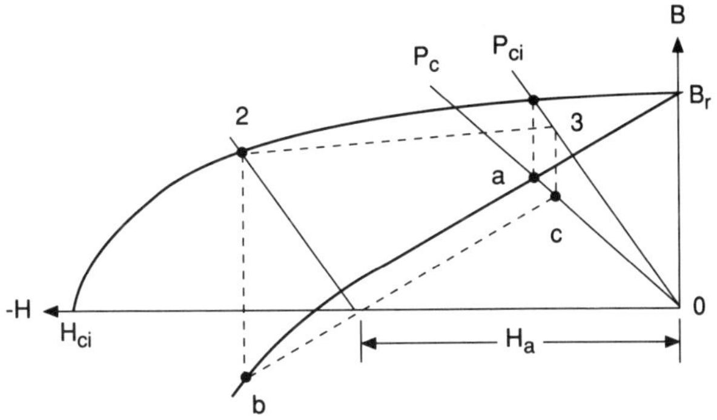

Figure 6.2 Effect of external demagnetising fields.

$$B_m = \mu_0 \frac{K_l A_g l_m}{K_r A_m l_g} H_m \tag{6.4}$$

For a given magnet and gap dimensions, equation (6.4) describes a straight line which intersects the demagnetisation curve at the working point ($-H_m$, B_m), as shown in Figure 6.1. This line is known as the *static load line*, and its slope, expressed by equation (6.3), is called the *permeance coefficient* (P_c).

The volume of the magnet material is found by rearranging equations (6.1) and (6.2) to give A_m and l_m, respectively. Upon multiplication we get:

$$A_m l_m = \frac{K_l K_r B_g^2 A_g l_g}{\mu_0 B_m H_m} \tag{6.5}$$

The term ($B_m H_m$) is known as the *energy product* of the magnet material, and the higher it is the lower the magnet volume will be.

The shortcoming of this simple design method lies in obtaining accurate values for the leakage and reluctance factors. This method will only give good predictions if all the leakage paths are taken into account, and if the level of saturation, and hence the reluctance, of the various soft iron sections is accurately determined.

6.1.2 Effect of demagnetising fields

Permanent magnets used in electric machines are normally subjected to demagnetising fields created by the other windings (armature reaction). The effect of a demagnetising force on a magnet is normally studied with the use of the intrinsic demagnetisation curve; as a demagnetising field would be acting against the magnet's inherent ability to have more flux with a given applied H than will appear in an air region. However, the corresponding change to the magnet's operating point on the normal demagnetisation curve must be observed; as this determines the reduction in flux produced by the magnet after the external field is removed.

The effect of a demagnetising force on a magnet is illustrated in Figure 6.2, which shows both the intrinsic and normal demagnetisation curves. The slope of the normal static load line is determined by the permeance coefficient (P_c), and the operating point on the normal demagnetisation curve is point (a). The intrinsic permeance coefficient (P_{ci}) is calculated from:

$$\frac{B_i}{H_m} = \frac{B_m + \mu_0 H_m}{H_m} = P_c + \mu_0 \tag{6.6}$$

the operating point on the intrinsic curve is at point (1), vertically above point (a) on the normal curve.

The operating point (2) on the intrinsic demagnetisation characteristic when a demagnetising field (H_a) is applied, lies on the line H_a–2 which is parallel to the intrinsic permeance line P_{ci}. The corresponding operating point on the normal curve is point (b), vertically under point (2). When the demagnetising field (H_a) is removed, the operating points, (2) and (b) on the intrinsic and normal curves, do not follow the demagnetisation curves back to points (1) and (a). Instead, they **recoil** on straight lines to points (3) and (c) on the intrinsic and normal permeance lines, respectively. The slope of the *recoil line* of each curve is approximately equal to the slope of the corresponding demagnetisation curve at B_r. The magnet is now said to be *stabilised*, and it will not be permanently affected by demagnetising fields equal to or less than H_a. The magnet now produces flux lines corresponding to its new operating point (c) and not point (a). Stabilisation is generally performed when Alnico magnets are used.

When ceramic and rare earth magnets are used, an efficient design is obtained when the armature reaction field shifts the magnet operating point to the knee of the intrinsic characteristic. Under such condition the operating point recoils on the major hysteresis curve and no partial demagnetisation occurs.

It should be noted that the coercive force temperature coefficient is positive for both ceramic and Alnico magnets (it is almost negligible for Alnicos). Therefore, the effect of external demagnetising fields must be studied at the lowest expected operating temperature. The opposite applies in the case of rare earth magnets.

6.2 PERMANENT MAGNET d.c. MOTORS

The use of permanent magnets to provide excitation in d.c. motors offer economic advantages in a certain size range. In miniature motors (under about 70 mm diameter) the wound-field structure can not compete with permanent magnets on cost basis. Above about 150 mm diameter economical considerations tend to favour wound-field motors. Obviously, this leaves an important size area in which the two types of motors compete on cost basis. Within this size range lie some of the automotive and most of the appliances motors.

The torque–speed characteristic of a permanent magnet d.c. motor is approximately a straight line drawn between the no load speed on the vertical axis and the stall torque on the horizontal axis. Due to the absence of field losses, permanent magnet motors are potentially more efficient than wound-field ones. Permanent magnet motors also have a broader range of high efficiency. The importance of high efficiency in increasing battery life is obvious for portable equipment.

Permanent magnet motors are not without their drawbacks. For example, when cost considerations dictate the use of ceramic magnets, the air-gap flux density will be lower than that of a wound-field motor of a comparable size. Although this would, in most cases, necessitate an increase in the armature diameter, it does not always result in a corresponding increase in the overall diameter. This is because the radial dimension of the magnet pole piece is always less than the wound pole height. The reason for this is easily explained by inspecting equation (4.30), chapter 4, which is reproduced here for convenience:

$$\frac{AT_f}{h_f} = \delta_{fc} b_{fc} J_{cf} \times 10^6 \qquad (6.7)$$

Using typical values for the parameters in the RHS of equation (6.7), the ampere-turns per metre of the field coil is found to be about 75 kA/m. The mmf of a typical ceramic magnet at the maximum energy product point is about 120 kA/m. Therefore, the magnet thickness would be about 60% of the field coil height.

Partial demagnetisation by armature reaction field may result if the permanent magnet motor is not properly designed. Nowadays, however, this does not present a serious problem because CAD techniques are available and they enable very accurate designs to be obtained.

6.2.1 Design of a permanent magnet motor

Mechanically commutated d.c. motors of the permanent magnet type may be classified into two main configurations; those in which the magnet pole face is adjacent to the armature and those in which the magnets are remote from the armature. Generally, the latter configuration exhibits higher leakage and therefore it requires a larger magnet volume. However, the remote magnet configuration enables focusing of the magnet flux, and in some designs it can yield air-gap flux densities higher than the magnet remanence.

In this book the first configuration is considered and the magnet design equations given above are utilised to determine the dimensions of the required magnet pole pieces. Thereafter, the design method will follow the general procedures described in chapter 4 for the design of wound-field motors.

Figure 6.3 illustrates the general configuration of a permanent magnet d.c. motor. It is shown that the magnet axial length is made, in the general case, longer than the armature. This results in a higher flux per pole being produced for a given stack length. The amount of flux increase is not directly proportional to the increase in magnet material and this may seem to be an inefficient way of increasing the pole flux. However, some magnet overhang is commonly used. This is primarily because it is practically difficult to precisely align the magnet pole pieces to the armature. The magnet overhang reduces the sensitivity of the armature flux to axial alignment. Additionally, magnet overhang and the resulting flux increase give rise to a smaller armature diameter being required. The reduction in armature iron and copper volume may result in lower overall motor cost.

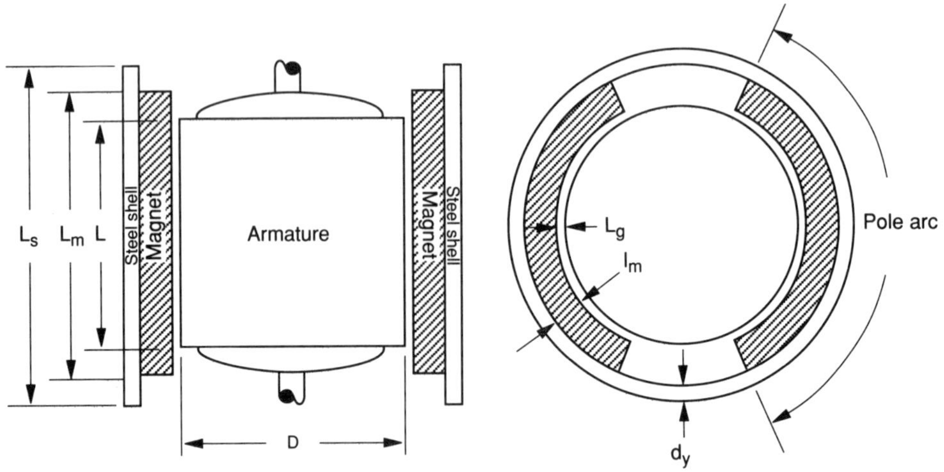

Figure 6.3 Permanent magnet d.c. motor configuration.

From experience, the leakage and reluctance factors for the permanent magnet d.c. motor configuration shown in Figure 6.3 are about 1.1 and 1.15 for normal designs, respectively. A 20% magnet overhang may be taken to increase the no load flux by about 15%. There are two common methods of fixing the magnet pole pieces to the outer steel shell which serves as the magnet's flux return path. The first utilises clips that mechanically fasten the magnets into the steel shell using spring pressure and is known as *clipped assembly*. The second method makes use of various bonding agents to bond the magnets to the shell surface and is known as *bonded assembly*.

Clipped assembly method has the advantage of being a fast, single step operation which lends itself to automation. The method is, however, restricted to 2-pole motors (as the variation in clip pressure makes 4-pole designs impractical) and to diameters of 75 mm or less (as the assembly provides a limited holding force).

With bonded assemblies there are no restrictions on the motor diameter or the number of poles. The bonded assembly also has a much higher impact resistance than the clipped assembly. On the other hand, the process involves surface preparation and application of the bonding material to the steel shell and/or the magnet pole pieces. Additionally, special devices may be required to hold the magnet in place while the bonding material cures.

6.2.2 Sample design

Design of a 50 W, 14 V, 6000 rev/min intermittent duty d.c. permanent magnet motor for automotive applications. The operating temperature range is from −20 °C to 100 °C.

The full load speed is 100 rev/s. Because of the high speed we design for two poles.

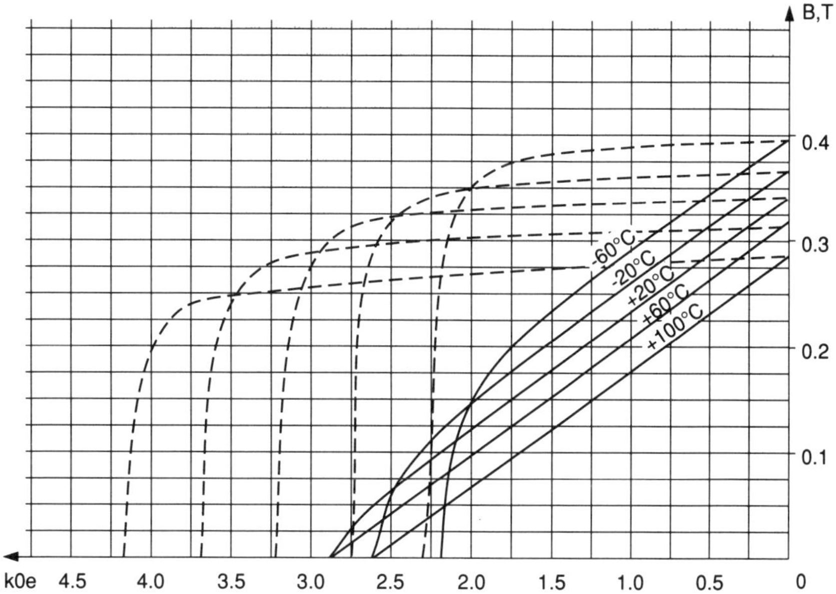

(1kOe = $\frac{10^3}{4\pi}$ kA/m)

Figure 6.4 Ceramic magnet demagnetisation characteristics.

Cost is of prime importance and a ceramic design is sought. The demagnetisation characteristics of a suitable grade of ceramic magnet material is shown in Figure 6.4.

In calculating the magnet flux we consider the characteristics at the highest operating temperature. Let the magnet operating flux density be 0.25 T at 100 °C. Therefore:

$$B_m = 0.25\,\text{T} \quad \text{and} \quad H_m = 26.3\,\text{kA/m}$$

The normal demagnetisation curve is used since we are determining the flux produced by the magnet but external to it.

Taking a pole arc to pole pitch ratio of 0.7, the required magnet arc is 126°.

Assuming a leakage factor of 1.15, which is typical of the configuration under consideration, and that a 20% magnet overhang increases the flux by 15%, the average air-gap flux density is calculated at 0.175 T. As the motor is intended for intermittent duty application, assume specific electric loading value of 8000 ampere-conductor per metre. The output coefficient is:

$$C_o = \pi^2 \times 0.175 \times 8000 \times 10^{-3} = 13.8$$

The D^2L product is:

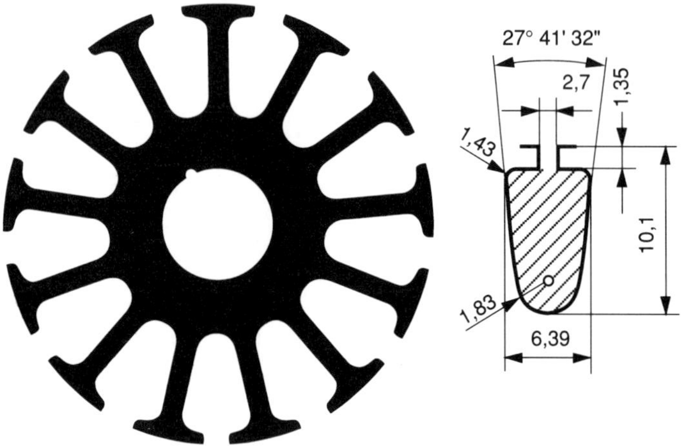

Figure 6.5 Armature lamination and slot details [6.1]. Dimensions are in mm.

$$D^2L = \frac{50 \times 10^{-3}}{13.8 \times 100} = 3.6 \times 10^{-5}\,\mathrm{m}^3$$

Taking the ratio of core length to pole pitch to be 0.9, the armature diameter and core length will be 39 mm and 24 mm, respectively.

The magnet axial length is taken as 29 mm and its radial length can only be determined when the air-gap length becomes known.

A small air-gap should be avoided because it stipulates tight tolerances for the magnet dimension, and this increases the magnets' cost. The air-gap radial length may be taken between 1% and 1.5% of the pole pitch or between 0.6 and 0.9 mm. Let the air-gap be 0.8 mm and assume an air-gap correction factor of 1.1. Taking the reluctance factor as 1.2, the magnet thickness is calculated as follows:

$$l_m = 1.2 \times \frac{B_g}{\mu_0} \times \frac{l_g}{H_m} = 1.2 \times \frac{0.25}{\mu_0} \times \frac{0.88}{26.3 \times 10^3} = 8\,\mathrm{mm}$$

The stacking factor is typically 0.925 and the average flux per pole at no load is 0.238 m Wb. Taking the no load speed to be 4% higher than the full load speed or 104 rev/s, the number of armature conductors is calculated as follows:

$$Z = \frac{14}{0.238 \times 10^{-3} \times 104} = 566$$

This number will be corrected to give an integer number of conductors per slot. The number of slots is selected so that the number of slots per pole is equal to a whole number plus $\frac{1}{2}$ in order to reduce flux pulsations. The 13-slot armature lamination shown in Figure 6.5 is selected [6.1]. With 44 conductors per slot the total number of armature conductors will be 572. The pole flux and the magnet operating point remain unchanged as the resulting difference is within the tolerance expected in material specification.

The efficiency is assumed at 0.75 p.u. and the armature current (or the motor terminal current) is:

$$I_a = \frac{50}{14 \times 0.75} = 4.8\ \text{A}$$

The conductor current is:

$$I_z = \frac{4.8}{2} = 2.4\ \text{A}$$

This is an intermittent duty motor and the current density is assumed 6 A/mm^2. The resulting conductor cross-section area is 0.4 mm^2. The nearest standard wire has a diameter of 0.71 mm and cross-section area of 0.396 mm^2. The total copper area of 44 conductors is 17.4 mm^2, giving a slot space factor of about 0.42. The length of the mean armature winding is:

$$L_{ma} = 2 \times 0.024 + 2.3 \times 0.061 + 5 \times 0.01 = 0.24\ \text{m}$$

and the armature resistance at 20 °C is:

$$R_a = \frac{572}{2 \times 2^2} \frac{1.77 \times 10^{-8} \times 0.24}{0.396 \times 10^{-6}} = 0.767\ \Omega$$

and at −20 °C:

$$R_a = 0.767(1 - 40 \times 3.82 \times 10^{-3}) = 0.65\ \Omega$$

The tooth width is 2 mm and the tooth flux density is:

$$B_t = \frac{2 \times 0.238 \times 10^{-3}}{13 \times 2 \times 10^{-3} \times 0.024 \times 0.925} = 0.82\ \text{T}$$

Although this is quite low, the tooth width at 2 mm is limited by mechanical considerations. Additionally, this value is at 100 °C, the pole flux increases as the temperature is reduced. The tooth flux density at −20 °C will be about 1 T.

Bearing in mind that the relative permeability of the ceramic magnet material is very close to unity and assuming that the magnet return path and armature lamination do not saturate as a result of the magnet and armature reaction field, the armature reaction demagnetising field is calculated from:

$$H_a = \psi \frac{I_{zp} Z}{2p} \frac{1}{(l_m + l_g)}$$

where I_{zp} is the peak conductor current which, obviously, occurs when the motor is stalled at the lowest operating temperature.

For the ceramic magnet to be protected against demagnetisation we must use the intrinsic characteristics corresponding to the lowest operating temperature or, in the present case, −20 °C. The conductor current when the motor is stalled at −20 °C is:

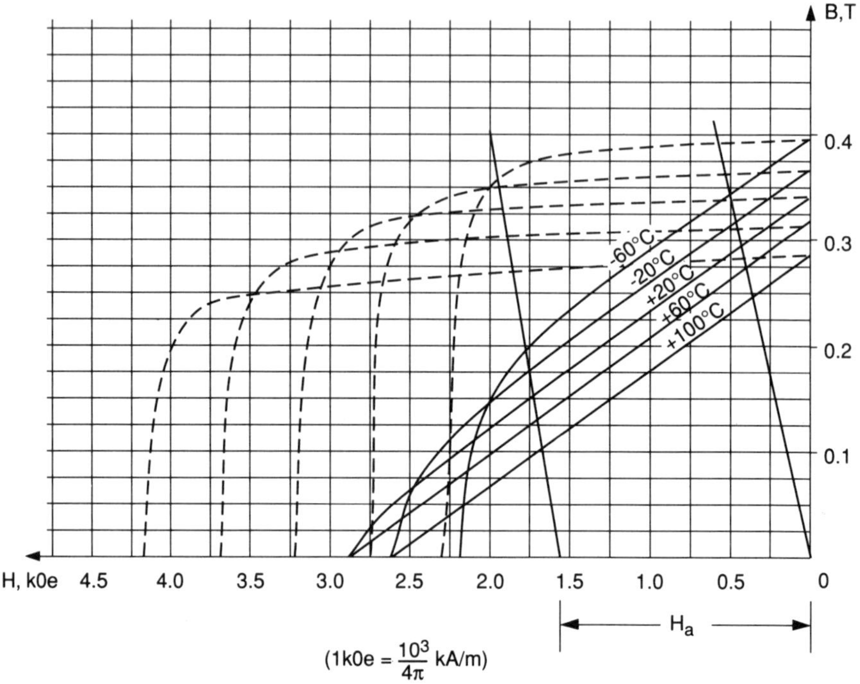

Figure 6.6 The effect of armature reaction on magnet operating point.

$$I_{zp} = \frac{14}{2 \times 0.65} = 10.77\,\text{A}$$

and:

$$H_a = 0.7\frac{10.77 \times 572}{2 \times 2}\,\frac{1}{8.88 \times 10^{-3}} = 121.4\,\text{kA/m}$$

The magnet operating point when the armature reaction field is at its strongest lies just before the knee of the intrinsic demagnetisation curve corresponding to $-20\,°C$, as shown in Figure 6.6. The magnet design is therefore satisfactory.

Having designed the permanent magnet excitation system, the design can be completed following the same procedure described in chapter 4 for the wound-field motors.

6.3 BRUSHLESS d.c. MOTORS

The idea of a brushless d.c. motor was first proposed over half a century ago, when thyratrons were the only available switching elements, and the level of technology was insufficient for this concept to be considered as anything other than of a passing interest. It is only in recent years (during the last two decades or so) that cost-effective practical systems have been built. The two most significant factors that led to the increasing popularity of this type of drive are improvements in magnet materials and the reduction in the cost of integrated circuits and electronic power switches. The recent development of high current, high voltage semiconductor switches has made it practical to construct such systems for various industrial applications.

A range of brushless d.c. drives are now in production and are becoming even more attractive with the falling cost, increasing simplicity and improved reliability of electronic components [6.2–6.6].

Brushless d.c. motors combine desirable characteristics of two conventional motors: they have the stall torque capability of a conventional brush-type d.c. motor and the high speed operation capability associated with a.c. induction or reluctance motor drives. Some of the advantages offered by brushless d.c. drive systems are summarised as follows:

(1) The elimination of the brush gear and commutator allows the drive motor to operate at much higher speeds than conventional d.c. machines, leading to reductions in weight and volume. Additionally, it leads to higher reliability and the motor requires, practically, no maintenance. The absence of commutator segments precludes the problem of segment oxidation. In some applications, brush arcing can be hazardous and a brushless motor is necessary for safety. Also, the lack of brushes eliminates certain types of radio frequency interference.

(2) The use of permanent magnet rotors leads to the elimination of rotor copper losses and considerable improvement in the thermal characteristics.

(3) The development of high energy permanent magnet materials such as samarium–cobalt and neodymium–iron–boron compounds allow the rotor diameter to be smaller than in a conventional brush-type motor. This yields lower rotor inertia and, consequently, faster acceleration.

6.3.1 System description

The main components of a brushless d.c. drive system are shown in the schematic block diagram of Figure 6.7. The drive motor has three elements: a fixed wound member or stator, a rotor carrying a permanent magnet excitation system, and a non-contacting means of sensing rotor position (rotor position sensors).

For a three-phase motor supplied from a six-step power conditioner, the operation of the system is as follows: the control logic receives information on the position of the rotor from the position sensors. This is then translated into one of six current states as shown in Figure 6.8. Each current state produces an armature electromotive force (emf) which remain fixed in space for 60 electrical degrees until the next current state. After the rotor completes 60 electrical degrees of rotation, the phase currents are, ideally, switched instantaneously into the next state. This results in a step of armature emf of 60 degrees in the direction of rotation. This process is repeated six times per electrical revolution resulting in a discretely jumping emf and continuous rotation.

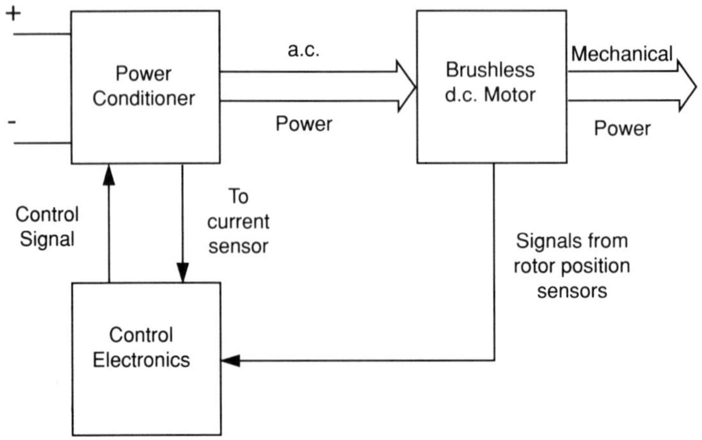

Figure 6.7 Schematic block diagram of brushless d.c. drives.

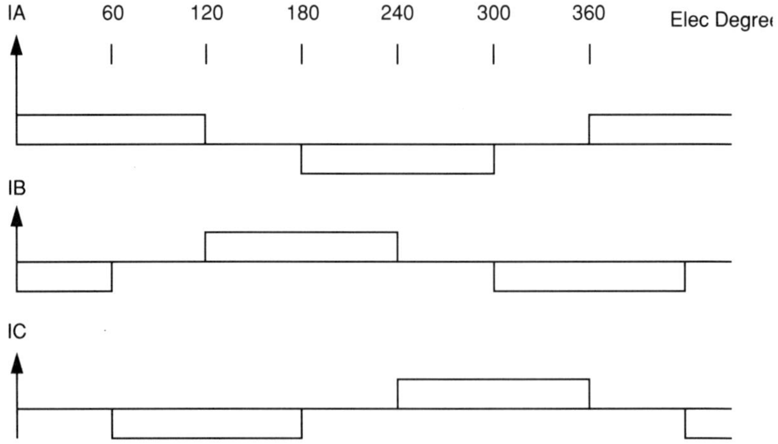

Figure 6.8 Idealised phase currents.

6.3.2 The motor

The majority of motors employed in brushless d.c. drives are three-phase units with the stator being a stationary outer winding (armature) and the rotor a rotational permanent magnet assembly (field).

Motors employed in high performance brushless d.c. drives utilise rare earth magnets while ceramic magnets are used in moderate performance, relatively inexpensive, drives. When ceramic magnets are used, the specific magnetic loading (or the air-gap flux density) will have a maximum value of about 0.2 T. Therefore, the stator (armature) should utilise narrow teeth and wide slots. On the other hand, in rare earth magnet designs the minimum magnet volume will be obtained when the magnet operating point is at about 0.55 T. Brushless d.c. motors utilising rare-earth magnets are rarely designed with an air-gap density value of less than about 0.5 T.

When the switching strategy is such that only two phases are excited at any one time, the effective number of armature conductors will be two thirds times the actual number of conductors; the specific electric loading is based on the effective number of conductors.

The stator (armature winding) is designed in a similar way as the primary winding of a polyphase induction motor. In fact, in many industrial applications it is possible to utilise an induction motor stator.

6.3.3 Position sensors

These sensors detect the position of the rotating magnets and send logic codes to a commutation decoder which, after processing this code, activates the firing circuits of semiconductor switches feeding power to the stator winding of the drive motor. A unidirectional torque is produced via the interaction between the permanent magnets and currents flowing through the winding. Reliable position sensing techniques do not involve contact between stationary and moving parts. Practical contactless sensors may be classified as follows:

- Optical sensors
- Reluctance sensors
- Magnetic field sensors
- Capacitance sensors
- Back emf sensors

A description of these techniques is given below.

Optical sensors

In these devices, a light source shines through a patterned disc attached to the rotor shaft and a photodiode detects the presence or absence of light. Both the light source and the photodiode are stationary.

The main advantage of this method is that the signal from the photodiode rises and falls quite abruptly and so the switching points are well defined. Also, the output signal is d.c., so it does not require rectification or filtering.

Disadvantages of optical sensors are high cost, and that they require a clean environment. Also, the light source is liable to fail suddenly. In addition, the photodiode signal is usually very weak and may require amplification before it can be used for control purposes.

Reluctance sensors

This method utilises a toothed magnetic wheel mounted on the rotor shaft and rotates between stationary poles carrying windings. A high frequency a.c. flux is directed around the magnetic circuit, and the amplitude of the signal induced in each pole winding, which is a function of the reluctance of the air-gap between the toothed wheel and the relevant pole, is monitored. Since the air-gap reluctance varies with the rotor position, it is possible to determine the rotor orientation from the amplitude of the induced signals.

The advantages of this method are low cost, simplicity of design, and suitability for use in harsh environments. The disadvantages include the need to rectify and filter the output signal and that the signal builds up gradually. Usually, a trigger circuit is needed to detect the correct switching points.

Magnetic field sensors

In this method the position of permanent magnets attached to the rotor is sensed by Hall effect sensors. The rotor magnetic field can be sensed directly, thus eliminating the need for any auxiliary magnets. The sensors must be located far enough from the stator winding to prevent the magnetic field generated by stator currents from interfering with their operation.

The advantage of magnetic field Hall sensors are freedom from RFI, suitability for a wide range of operating conditions, and high accuracy. In the past, engineers were reluctant to use the Hall effect sensors because the cost of Hall hybrids and integrated circuits was much higher than that of optical or mechanical components associated with the reluctance sensors. The cost of Hall components has dropped significantly in recent years and now they are widely used in various applications.

Capacitance sensors

These devices are somewhat similar to the reluctance sensors. The position of a suitably shaped wheel mounted on the rotor shaft is detected by a capacitance measuring probe. Capacitance sensors are compact and can be easily incorporated within the motor frame. However, considerable amplification is needed to boost the output signal to a useable level.

Back emf sensors

It is possible to detect the rotor position by monitoring the back emf waveforms generated in the stator winding by the rotor flux [6.7]. This technique eliminates the labour cost associated with mechanical adaptation and maintenance of position sensors inside the motor, and, therefore, it improves reliability. The disadvantages are the effect of generated noise on the signals due to the switching action in the winding, which can make accurate position detection difficult. Additionally, the rotor position can not be determined at very low speeds since the back emf values are very small.

Table 6.1 Design trade-offs for three types of sensors

Parameter	Mechanical	Optical	Hall
Life expectancy	3	2	1
Design complexity	1	2	3
RFI (sparking)	3	1	1
Cost	1	3	3
Range of environment	1	3	1
Accuracy	3	1	1

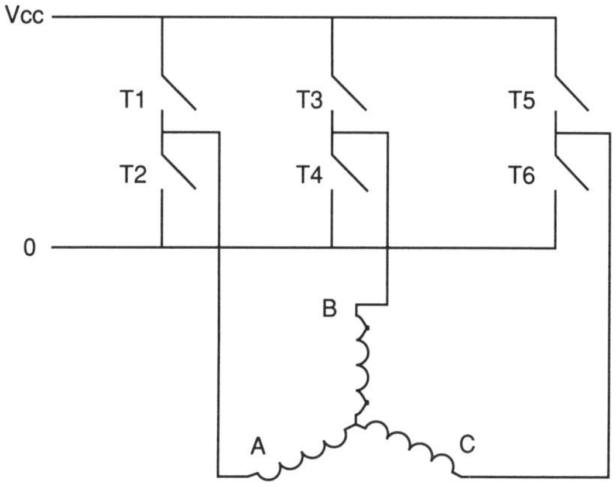

Figure 6.9 An illustration of the principle of brushless d.c. motors.

Comparison between position sensing techniques

Of the position sensors discussed above, the Hall effect, reluctance, and optical sensors are more practical. The design trade-offs of these sensors are given in Table 6.1, where 1 is best and 3 is worst [6.8].

6.3.4 Commutation decoder

The commutation decoder receives signals from the position sensors regarding the position of the rotor and translates them into switching signals supplied to the firing circuits of the electronic switches. In a six-step drive system, two switches are activated simultaneously to energise two of the stator phases, and a rotating field is established in the air-gap by the combining effect of the excitation current of the two coils.

The general principle of a three-phase brushless d.c. motor drive is illustrated in Figure 6.9, using the more widely applied six switch circuit as an example.

With reference to Figure 6.9, switches T1 and T4, T1 and T6, T3 and T6, T3 and T2, etc. are activated sequentially to move the field around the air-gap in the forward (clockwise) direction. The switching sequence and the direction of the

Coils Energised	A & B	A & C	B & C	B & A	C & A	C & B
Direction of Magnetic Field	30°		60°	60°		30°
Switches Activated	T1 & T4	T1 & T6	T3 & T6	T3 & T2	T5 & T2	T5 & T4
Shaft Angle	0-60	60-120	120-180	180-240	240-300	300-360

Figure 6.10 Switching sequence and resulting air-gap field direction.

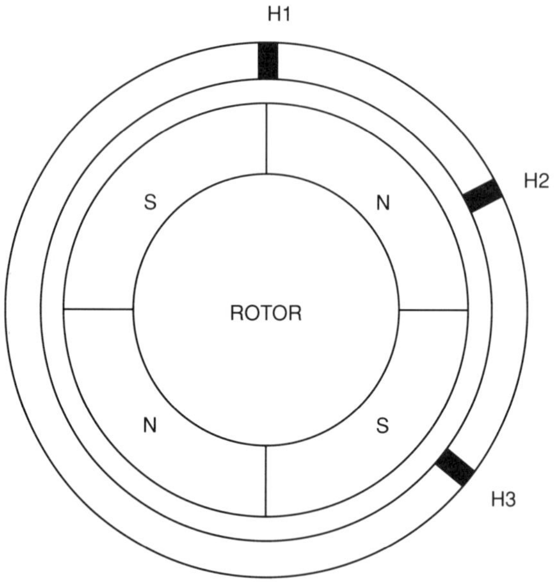

Figure 6.11 Four-pole permanent magnet rotor and three sensors 60° apart.

resulting air-gap field are shown in Figure 6.10 for forward (clockwise) rotation. For reverse (anticlockwise) rotation, the switching sequence is reversed.

In general, for a system employing three position sensors and a six-switch inverter bridge, a 3-to-6 line decoder is needed. The three input lines are connected to the output of the sensors, and the six outputs are connected to the switches firing circuits. The sensors may be arranged with 30, 60, or 120 mechanical degrees spacing. The decoder design is discussed here first for the case of 60 degrees spacings, and then the discussion is extended to cover the other two configurations.

A schematic of the rotor cross-section and the position sensors is shown in Figure 6.11. Hall sensors are assumed and each sensor operates when it is passed by a south pole and released when it is passed by a north pole. For the initial position shown in Figure 6.11, the first sensor (H1) operates at 0°, the second (H2) operates after 60° and the third (H3) releases after 30° of rotation in the clockwise (forward)

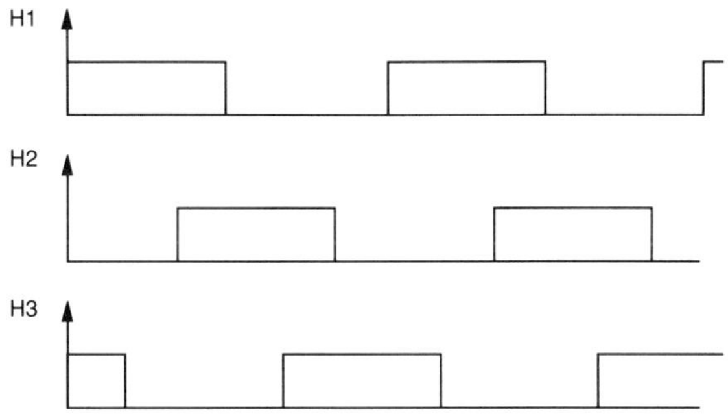

H1	1	1	1	0	0	0	1	1	1	0	0	0
H2	0	0	1	1	1	0	0	0	1	1	1	0
H3	1	0	0	0	1	1	1	0	0	0	1	1

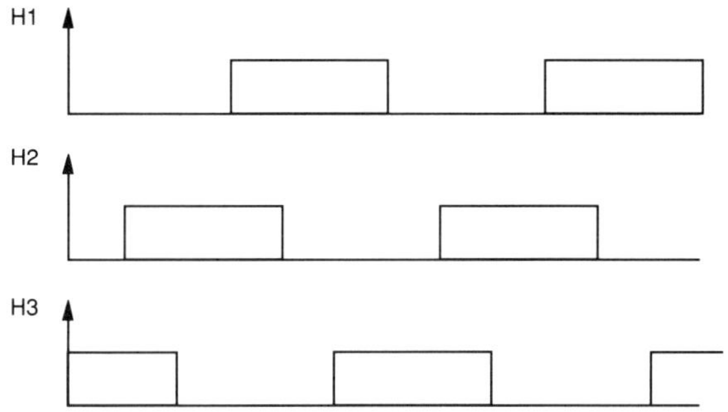

H1	0	0	0	1	1	1	0	0	0	1	1	1
H2	0	1	1	1	0	0	0	1	1	1	0	0
H3	1	1	0	0	0	1	1	1	0	0	0	1

Figure 6.12 Sensors output and corresponding logic codes.

direction. Figure 6.12 shows the sensors output code for forward and reverse rotation. It shows that there are six different commutation logic combinations per one electrical revolution and that the logic codes for both directions of rotation are the same, but the code sequence is reversed.

Similarly, the logic codes for the cases of 30° and 120° spacing between sensors are obtained. These are shown in Table 6.2 for forward rotation.

It should be noted that whether the sensors are placed 30, 60 or 120 mechanical degrees apart, they always resolve the rotor position to 60 degrees of electrical revolution. What distinguishes the three placements is the particular coding for each 60 degrees increment of rotor position.

The logic values of H2 and H3 with 120° placement are the same logic values of H3 and H2 with 60° placement respectively. Thus, by swapping H2 and H3 inputs, a decoder designed for one configuration can be used for the other. Similarly, the

Table 6.2 Hall sensors logic codes

Sensors phasing (mechanical degrees)	Position range (electrical degrees)	Sensors output codes		
		H1	H2	H3
30°	0–60	0	0	0
	60–120	0	0	1
	120–180	0	1	1
	180–240	1	1	1
	240–300	1	1	0
	300–360	1	0	0
60°	0–60	1	0	1
	60–120	1	0	0
	120–180	1	1	0
	180–240	0	1	0
	240–300	0	1	1
	300–360	0	0	1
120°	0–60	1	1	0
	60–120	1	0	0
	120–180	1	0	1
	180–240	0	0	1
	240–300	0	1	1
	300–360	0	1	0

output logic values of H1 and H3 with 30° placement are the inverse of the values of H1 and H3 with 60° placement respectively. By using two exclusive OR gates (XOR) at H1 and H3 decoder inputs, a decoder designed for the 60° placement configuration can be used when the phasing between the sensors is 30°.

The decoder truth table is readily derived with the aid of Figure 6.10 and Table 6.2. In order to enable rotation in both directions, a code corresponding to the desired direction of rotation (F/R select) is added as an input to the commutation decoder. Here codes 0 and 1 are chosen for forward (clockwise) and backward rotation respectively, but this is arbitrary. Other inputs to the commutation decoder consist of the position sensors output codes. The commutation decoder truth table is given in Table 6.3 for the case of three position sensors with 60° spacing. The decoder outputs determine the required state of the power conditioner switches for forward and reverse rotation, where codes 0 and 1 correspond to switch off and on respectively.

The Boolean functions of the semiconductor switches are derived in Appendix 4. They are as follows:

$$T1 = \overline{\overline{(\overline{F}\,.\,H1\,.\,\overline{H2})}\;\overline{(R\,.\,\overline{H1}\,.\,H2)}} \tag{6.8}$$

$$T2 = \overline{\overline{(\overline{F}\,.\,\overline{H1}\,.\,H2)}\;\overline{(R\,.\,H1\,.\,\overline{H2})}} \tag{6.9}$$

$$T3 = \overline{\overline{(\overline{F}\,.\,H2\,.\,\overline{H3})}\;\overline{(R\,.\,\overline{H2}\,.\,H3)}} \tag{6.10}$$

Table 6.3 Truth table for forward and reverse rotation

Direction of rotation	Rotor position range	Decoder input				State of switches					
		F/R	H1	H2	H3	T1	T2	T3	T4	T5	T6
Forward	0–60	0	1	0	1	1	0	0	1	0	0
	60–120	0	1	0	0	1	0	0	0	0	1
	120–180	0	1	1	0	0	0	1	0	0	1
	180–240	0	0	1	0	0	1	1	0	0	0
	240–300	0	0	1	1	0	1	0	0	1	0
	300–360	0	0	0	1	0	0	0	1	1	0
Reverse	0–60	1	0	0	1	0	0	1	0	0	1
	60–120	1	0	1	1	1	0	0	0	0	1
	120–180	1	0	1	0	1	0	0	1	0	0
	180–240	1	1	1	0	0	0	0	1	1	0
	240–300	1	1	0	0	0	1	0	0	1	0
	300–360	1	1	0	1	0	1	1	0	0	0

$$T4 = \overline{\overline{(\overline{F}\,.\,\overline{H2}\,.\,H3)}\;\overline{(R\,.\,H2\,.\,\overline{H3})}} \tag{6.11}$$

$$T5 = \overline{\overline{(\overline{F}\,.\,H3\,.\,\overline{H1})}\;\overline{(R\,.\,\overline{H3}\,.\,H1)}} \tag{6.12}$$

$$T6 = \overline{\overline{(\overline{F}\,.\,\overline{H3}\,.\,H1)}\;\overline{(R\,.\,H3\,.\,\overline{H1})}} \tag{6.13}$$

Commutation decoder using logic gates

The six Boolean functions describing the operation of the power conditioner switches are used to construct the logic diagram of Figure 6.13. In order to minimise the propagation delay through the circuit, the Boolean functions are implemented by NAND gates rather than AND/OR gates. Each function is implemented using two 3-input NAND gates and one 2-input NAND gate.

Commutation decoder using an EPROM

Once programmed, an EPROM holds its stored data indefinitely. The decoder truth table (Table 6.3) is used to program the EPROM. The input data (i.e. three outputs of the position sensors and the command indicating the desired direction of rotation) are used to define the EPROM memory addresses. At these locations, data from the output of the truth table, relating to the state of the semiconductor switches, are stored. Additionally, protection functions, such as over-current and over-voltage protection, can easily be incorporated.

Compared with the logic gates decoder, the EPROM offers the advantage of ease of incorporating user defined functions. Also, once the software has been written, it can be used to program a large number of EPROMs.

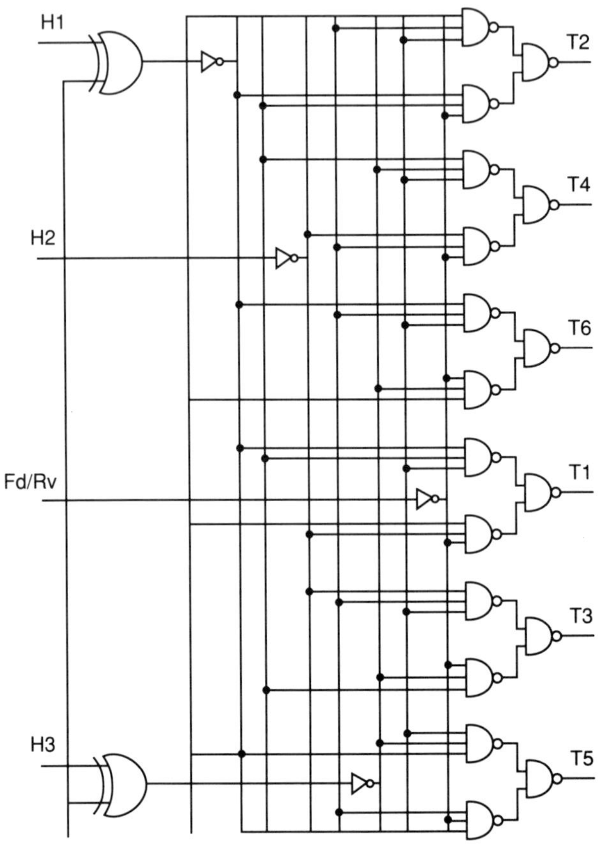

Figure 6.13 Commutation decoder logic diagram.

6.3.5 Switching elements

The application requirements influence the choice of the switching elements to be used in the power conditioning unit. The bipolar and MOSFET power transistors switch considerably faster than thyristors and GTOs. Typically, for a sub-kW system, switching times would be in the order of 1 to 2 μs for bipolar transistors and 30 to 150 ns for MOSFETs, and can be controlled by the base current and gate voltage respectively.

The gate of a thyristor loses control after turn-on is achieved. To turn off a thyristor its current must be interrupted for a short time. This is normally done, in d.c. circuits, by using a forced commutation circuit. To turn off a GTO, it is necessary to provide reverse gate current which is maintained during the whole recovery period.

The cost of additional turn off circuitry required for thyristors and GTOs is greater than the total cost of equivalent schemes employing bipolar or MOSFET transistors. On the other hand, the transistors cannot compete with any device in the thyristor family in terms of voltage and current handling capability. However, small brushless d.c. motors are not particularly demanding in terms of device rating.

The power MOSFET offers several unique features that make it an attractive device for use in small converters operating at high switching frequency. These features are summarised in the following points [6.9]:

(1) It has considerably faster switching speeds, and therefore, low switching losses.

(2) It is a voltage-driven device with extremely high input impedance, hence it requires simpler gate drive circuitry and much lower driving power.

(3) It has an integral body-drain diode, which, in voltage inverters, serves as a freewheeling diode, thereby eliminating the need for an external one.

(4) It is suitable for parallel operation. This is because it has positive temperature coefficient which forces the current to be uniformly distributed between the paralleled devices.

On the other hand, the power MOSFET has a relatively low power rating and high conduction losses, which increase rapidly as the device voltage rating is increased. However, high power rating and low conduction losses can be achieved by paralleling several low voltage power MOSFETs, and, depending on the application requirements, may be cheaper than an equivalent high power rating device.

6.4 LINE-START PERMANENT MAGNET MOTORS

Line-start permanent-magnet machines are high efficiency constant-speed polyphase motors, designed mainly to operate from a fixed frequency and fixed voltage power supply. These motors offer a very high efficiency alternative to induction motors which are used in the vast majority of industrial, commercial, and consumer applications today. Their high efficiency is due to the use of permanent magnets to provide rotor excitation instead of electromagnets; which inherently have sizeable resistive losses during normal operation. As well as the permanent magnet excitation system, the rotor of such machines houses a cage winding to provide the necessary start-up and accelerating torque. The machine utilises a conventional multi-phase stator winding.

6.4.1 Operation

When connected to a balanced a.c. supply, the motor runs-up towards synchronous speed, the torque being produced by the cage winding, and then attains synchronous speed through the process of synchronisation.

The starting torque is developed via interaction between the rotating magnetic field, which is established in the air-gap by the stator excitation, and currents induced in the cage winding. As the rotor starts to rotate, the permanent magnet

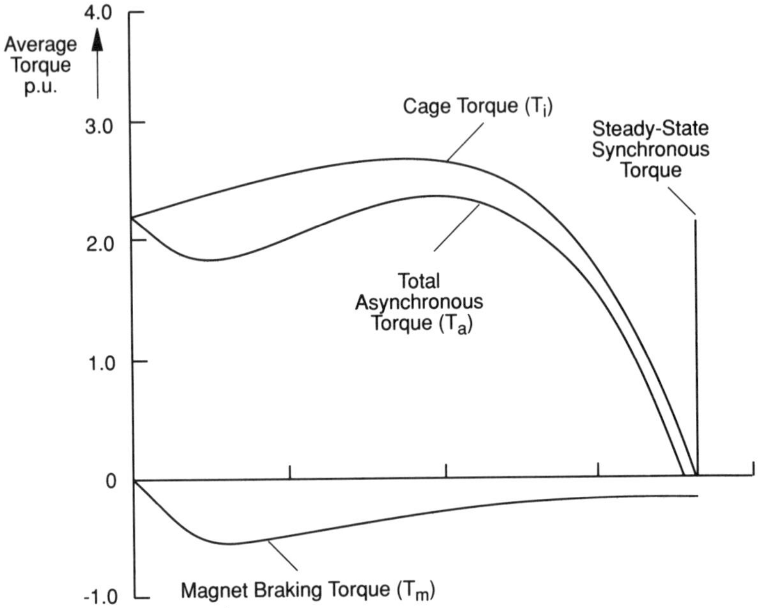

Figure 6.14 Typical variation, with speed, of asynchronous torque components.

flux induces slip-frequency emf and current in the stator winding. The torque associated with this current (magnet braking torque) opposes the induction accelerating torque during run-up. The nett torque (T_a) which accelerates the rotor towards synchronous speed is therefore the difference between the induction torque (T_i) and the magnet braking torque (T_m), as illustrated in Figure 6.14. After synchronisation, the induction torque (T_i) vanishes and the magnet torque (T_m) reverses its role and becomes the sole source of torque.

6.4.2 Review of some designs

Although the increased interest in permanent magnets occurred with the introduction of Alnico alloys and ceramic magnets, it was the development of the high energy rare earth magnets that revolutionised their use in electric motors. Many different configurations of permanent magnet synchronous motors, and generators, have emerged in recent years. Some of these designs are described below.

The "Permasyn" motor [6.10–6.13]

This is believed to be the earliest "modern" design of a permanent magnet synchronous motor. It is due to Merrill, and was reported in 1955 [6.10]. The rotor of this machine consists of an assembly of magnets inside a thin, soft laminated

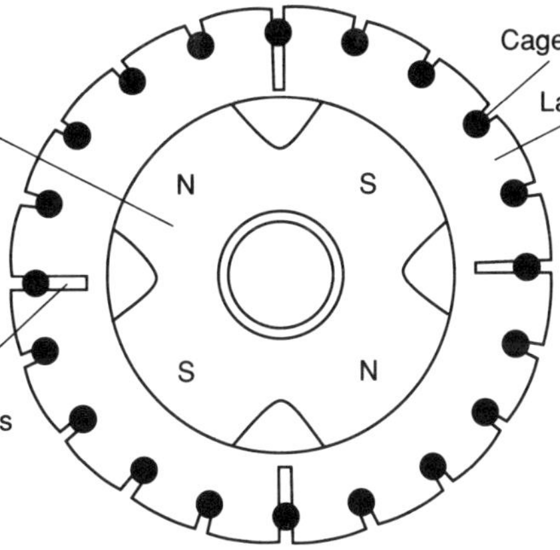

Figure 6.15 The "Permasyn" rotor configuration [6.10].

steel ring, thus producing poles on the rotor surface, as shown in Figure 6.15. The magnet material used in this work was cast Alnico. The steel ring housed a cage winding near the air-gap surface, and it was supposed to saturate to prevent short-circuiting of the magnets. Two rotors, the first a 4-pole and the second a 2-pole design, were built and tested. The results demonstrated that the general idea was sound, and warranted further development.

Due to its inherent design problems and performance limitations, the "Permasyn" motor did not attract much commercial interest. However, judging by the surge of interest that followed the publication of Merrill's work, it can be said that this work has served to identify the potential of this type of motors.

Douglas [6.11, 6.12], and Cahill and Adkins [6.13] followed up Merrill's work by developing the theory of operation and design methods for the "Permasyn" motor. Their work improved the understanding of the operation of these machines and identified how their performances can be improved.

The "Siemosyn" motor [6.14]

Introduced by Siemens, this motor has a somewhat complicated construction. The rotor utilises permanent magnet segments assembled within the core (which is made up from three laminated sections separated by non-magnetic material), and the whole assembly is held together by a die-cast cage as shown in Figure 6.16. The earlier designs utilised Alnico magnets, but rare earth materials were used in later designs.

The main disadvantages of this configuration are: the complicated structure, and unsuitability of applications requiring high starting performance [6.15]. The latter is because the flow of the induced alternating flux is obstructed by the presence of the magnet segments in the rotor magnetic circuit; leading to only a part of the cage winding being effective in producing induction motor action.

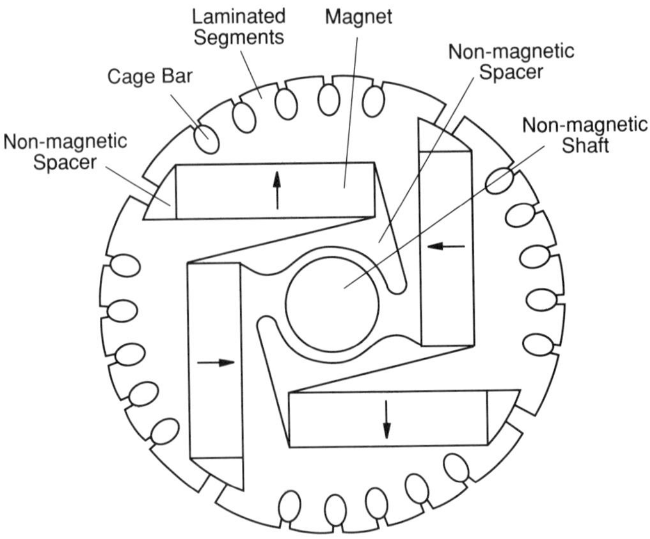

Figure 6.16 The "Siemosyn" rotor configuration [6.14].

The "Binns" configurations [6.15–6.23]

A valuable contribution to the field of permanent magnet synchronous motors is due to Binns and his co-researchers. Their configurations may be regarded as two different designs; both of the buried magnet rotor type.

The first configuration [6.15–6.19] utilises a single set of rotor laminations punched to accommodate magnet pole pieces and flux barriers. The magnets were secured by non-magnetic wedges and the whole structure is strengthened by the introduction of the cage winding. The effect of concentrating the air-gap flux over a small rotor pole arc on the motor's performance was examined and reported to enhance the pull-out capability and synchronising performance of these machines [6.19]. However, this design suffers from the drawback of constructional complexity, as is evident from Figure 6.17.

The second design has a more intricate construction than the first [6.20–6.23]. However, this design produced higher air-gap field than the first configuration. This is due to magnet polarisation over the whole pole arc, as shown in Figure 6.18. This configuration is rather similar to the earlier Siemens design described above; the difference being the relative disposition of the magnets within the rotor. Unlike Binns's first design, this second configuration exhibits very little reluctance action.

Ceramic magnets made of barium or strontium-ferrite were used in earlier designs, but later experimental machines utilised high energy rare earth magnets [6.22, 6.23]. This type of hybrid reluctance permanent magnet motor is reported to yield good performance, but it requires special production techniques [6.23]. These designs, however, have not received wide industrial acceptance. This is mainly due to the complexity of their construction.

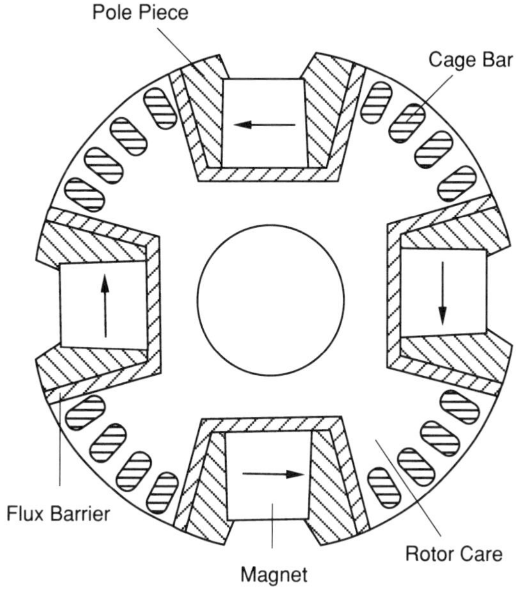

Figure 6.17 The first "Binns" configuration [6.15].

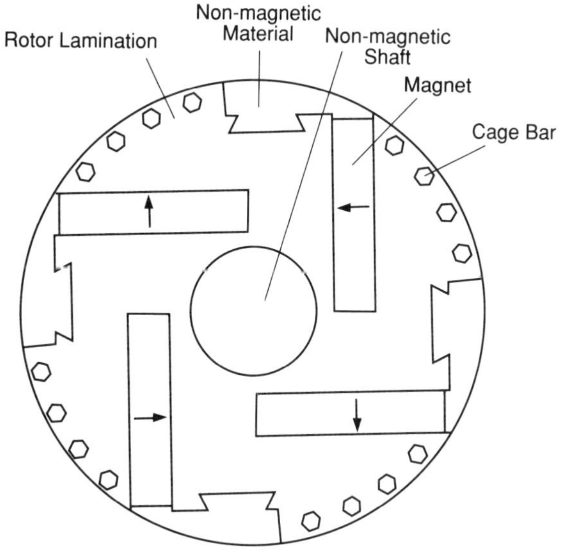

Figure 6.18 The second "Binns" configuration [6.20].

The "Isosyn" motor [6.24–6.26]

This configuration was introduced by Bermond and Flynn [6.24], and was subsequently analysed and evaluated by Chalmers and his co-researchers at UMIST [6.25, 6.26]. The rotor of this machine comprises laminations built up in a sectionalised stack (to allow skewing of the rotor). It utilises radially-oriented

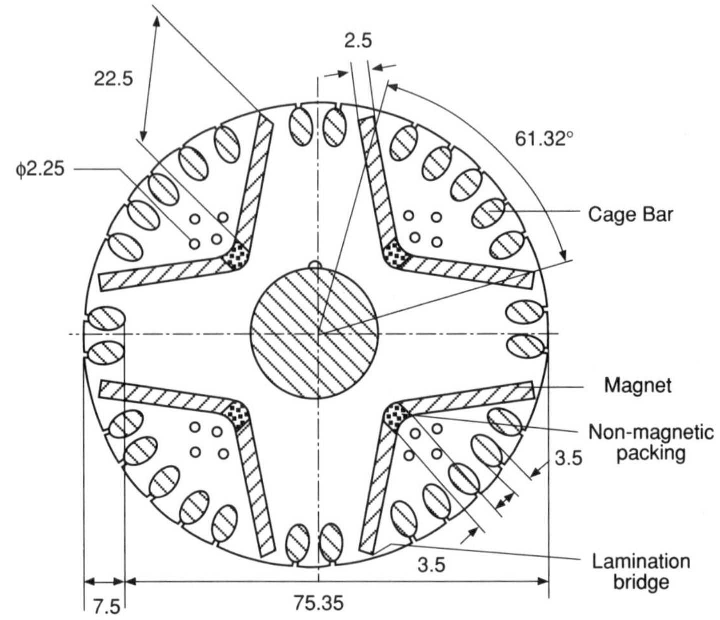

Figure 6.19 The "Isosyn" rotor configuration [6.25].

samarium–cobalt magnet segments housed in V-shaped slots; in order to maximise effective magnet cross-section area within the rotor. A cage winding is provided to produce induction torque for line-start applications. Small power Isosyn motors have been reported to be already employed on railway stock to drive fans [6.24]. Figure 6.19 shows the rotor of this design [6.26].

The GE design [6.27–6.31]

The rotor of this design comprises laminated pole pieces sandwiching permanent magnet segments between themselves and the main rotor body. The pole pieces house a cage winding and they are held in position by interpolar non-magnetic wedges, which are bolted to the core, as illustrated in Figure 6.20 [6.27, 6.28]. Additional conducting bars are provided in the wedges to obtain a balanced cage distribution. Samarium–cobalt magnets were used in the rotor of a 25 hp, 60 Hz, 4-pole motor. For high duty applications, it is estimated that the magnet cost could be recovered in energy cost savings in two to three years [6.28].

Trade-off studies aimed at developing the GE motor for industrial applications for both ferrites and rare earth magnet materials was reported [6.30]. The emphasis was on maximum utilisation of magnet materials. A most interesting conclusion of this work is that for performance specification equivalent to that of high efficiency induction motors, synchronous motors based on either material do not require any larger frame size than the comparable induction motor up to ratings of 18.6 kW (25 hp) for 4-pole machines [6.30].

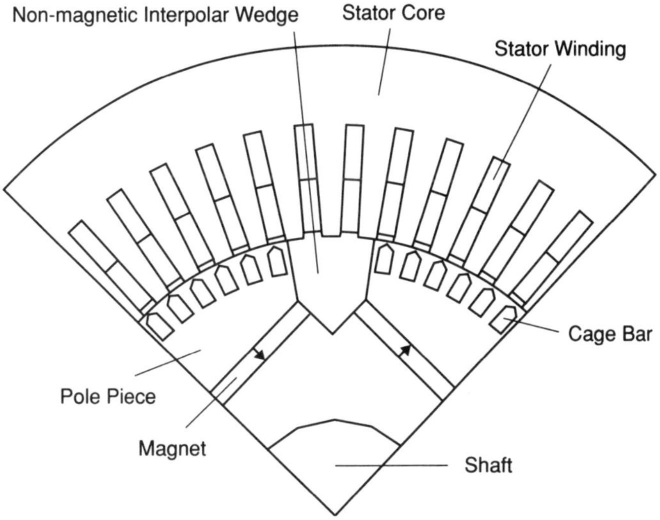

Figure 6.20 The GE rotor design [6.27].

Westinghouse designs [6.32, 6.33]

An earlier configuration, utilising Alnico or ceramic magnet materials, was introduced by the Westinghouse Corporation in 1969 [6.32]. The design and construction were similar in principle to the "Permasyn" [6.10]. The main disadvantage of this earlier design is that, like the "Permasyn", the magnets were exposed to demagnetisation.

The introduction of Westinghouse's most recent design coincided with the development of neodymium–iron–boron magnets [6.33]. This construction is rather similar to the GE design discussed above [6.27–6.31]. The most obvious difference is that the Westinghouse construction requires only one set of rotor laminations in which slots are provided to accommodate magnet segments as well as a conventional cage winding; as shown in Figure 6.21. High energy magnets are used in this design, which reduces the risk of demagnetisation.

6.5 REFERENCES

[6.1] Kienle and Spiess (1985) *Armature Laminations*, Stanz- und Druckgießwerk GmbH, D-7123 Sachsenheim, Germany.

[6.2] Nicholson, D. S. (1991) *Drives/Motors/Controls Conf.*, Birmingham, England, p. 3.1.

[6.3] Kenjo, T. and Nagamori, S. (1980) *Permanent-magnet and Brushless d.c. Motors*, Clarendon Press, Oxford.

[6.4] Artusi, D. and Hattori, H. (1986) *Proc. IECON'86*, Milwaukee, WI, USA, p. 772.

[6.5] Silva, C. (1985) *Electronic Design*, Sept., p. 149.

[6.6] Moyes, A. J. and Murrell, A. E. (1985) *IEEE Conf. on Drives, Motors and Controls*, London, England, p. 107.

[6.7] Binns, K. J. *et al.* (1990) *IEE Proc.*, **137**, Pt. B, No. 4, p. 253.

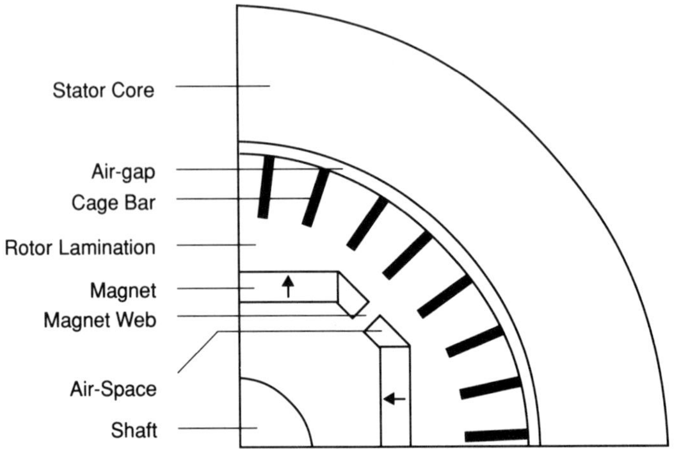

Figure 6.21 The Westinghouse second rotor design [6.33].

[6.8] Lundin, W. F. (1985) *Proc. of 7th Int. MOTOR-CON Conf.*, Hannover, Germany, p. 126.

[6.9] Wood, P. (1985) *HEXFET Data Book*, Application note No. 950-A.

[6.10] Merrill, F. W. (1955) *AIEE Trans.*, **74**, Pt. 3, p. 1754.

[6.11] Douglas, J. F. H. (1959) *AIEE Trans.*, **78**, Pt. 3, p. 76.

[6.12] Douglas, J. F. H. (1959) *AIEE Trans.*, **78**, Pt. 3, p. 221.

[6.13] Cahill, D. B. M. and Adkins, B. (1962) *IEE Proc.*, Pt. A, **109**, No. 48, p. 483.

[6.14] Siemens (1970) An electric machine having permanent magnets mounted in the rotor between its pole segments, UK Patent No. 1177247.

[6.15] Binns, K. J. and Barnard, W. R. (1971) *IEE Proc.*, **118**, No. 2, p. 369.

[6.16] Binns, K. J. *et al*: (1976) *IEE Conf. on Small Electrical Machines*, London, England, p. 76.

[6.17] Binns, K. J. and Jabbar, M. A. (1976) *IEE Conf. on Small Electrical Machines* London, England, p. 78.

[6.18] Binns, K. J. *et al.* (1978) *IEE Proc.*, **125**, No. 3, p. 203.

[6.19] Binns, K. J. *et al.* (1979) *IEE Proc.*, **126**, No. 8, p. 741.

[6.20] Binns, K. J. and Jabbar, M. A. (1980) *ICEM*, Athens, Greece, p. 304.

[6.21] Binns, K. J. and Jabbar, M. A. (1981) *IEE Conf. on Small Electrical Machines*, London, England, p. 107.

[6.22] Binns, K. J. and Jabbar, M. A. (1981) *IEE Proc.*, **128**, Pt. B, No. 3, p. 157.

[6.23] Binns, K. J. and Wong, T. M. (1984) *ICEM*, Lausanne, p. 595.

[6.24] Bermond, G. and Flynn, J. B. (1981) *IEE Conf. Publ. No. 202*, p. 119.

[6.25] Chalmers, B. J. *et al.* (1985) *IEE Proc.*, **132**, Pt. B, No. 3, p. 117.

[6.26] Chalmers, B. J. *et al.* (1986) *ICEM*, Munich, Germany, p. 1067.

[6.27] Miller, T. J. E. *et al.* (1983) *Conf. Record of IEEE/IAS Annual Meeting*, Mexico City, p. 455.

[6.28] Cornell, E. P. (1983) *Drives, Motors, Controls Conf.*, Leeds, England, p. 102.

[6.29] Miller, T. J. E. (1984) *IEEE Trans.*, **PAS-103**, No. 7, p. 1822.

[6.30] Richter, E. and Neumann, T. W. (1984) *IEEE Trans.*, **MAG-30**, No. 5, p. 1762.

[6.31] Richter, E. *et al.* (1985) *IEEE Trans.*, **IA-21**, No. 4, p. 644.

[6.32] Westinghouse (1969) Permanent magnet rotor, UK Patent No. 1147917.

[6.33] Ortoli, E. S. *et al.* (1987) *9th Int. Workshop on Rare-Earth Magnets and Their Applications*, Bad-Soden, Germany, Paper No. W1.4, pp. 41–53.

6.6 BIBLIOGRAPHY

Binns, K. J. and Kurdali, A. (1979) 'Permanent-magnet ac generators', *Proc. IEE*, **126**, No. 7, p. 690, July.

Binns, K. J. *et al.* (1982) 'Some aspects of self-starting permanent magnet machines', *Industrial Conf. on Electrical Machines (EMDA)*, London, England, p. 236.

Binns, K. J. and Low, T. S. (1983) 'Performance and application of multi-stacked imbricated permanent magnet generators', *Proc. IEE*, **130**, Pt. B, No. 6, p. 407, November.

Binns, K. J. and Wong, T. M. (1984) 'Analysis and performance of a high-field permanent-magnet synchronous machine', *Proc. IEE*, **131**, Pt. B, No. 6, p. 252, November.

Chen, M. and Levi, E. (1985) 'Iron saturation in PM ac Motors', *IEEE Trans.*, **MAG-21**, No. 3, p. 1262, May.

Cheng, M. and Zhou, E. (1992) 'Pull-in criterion for permanent-magnet synchronous motors', *ICEM*, Manchester, England, p. 790.

Consoli, A. and Abela, A. (1986) 'Transient performance of permanent magnet ac motor drives', *IEEE Trans.*, **IA-22**, No. 1, p. 32.

Consoli, A. and Renna, G. (1989) 'Interior type permanent magnet motor analysis by equivalent circuits', *IEEE Trans. on Energy Conf.*, Vol. 4, No. 4, p. 681, December.

Demerdash, N. A. and Nehl, T. W. (1978) 'Numerical simulation of dynamics of brushless d.c. motors for aerospace and other applications', Final Report on Virginia Polytechnic Institute and State University—NASA (JSC) Contract No. NAS9-15091.

Evans, P. D. and Brown, D. (1990) 'Simulation of brushless d.c. drives', *IEE Proc.*, **137**, Pt. B, No. 5, p. 299.

Hall, J. K. *et al.* (1984) 'Switching properties of gate turn-off thyristors', IEE Conf. on Power Electronics and Variable Speed Drives, 1–4 May, *IEE Conf. Publ. No. 234*, p. 58.

Hanrahan, D. J and Toffolo, D. S. (1957) 'Permanent magnet generators, Part I—Theory', *AIEE Trans.*, **76**, Pt. 3, p. 1098.

Honsinger, V. B. (1980) 'Permanent magnet machines—asynchronous performance', *IEEE Trans.*, **PAS-99**, No. 4, p. 1503.

Honsinger, V. B. (182) 'The fields and parameters of interior type ac permanent magnet machines', *IEEE Trans.*, **PAS-101**, No. 4, p. 867, April.

Huffman, H. H. (1990) 'Introduction to solid-state adjustable speed drives, *IEEE Trans. on Industry Applications*, **26**, No. 4, July/August, p. 671.

Hui, S. Y. R. and Bolton, H. R. (1988) 'Microprocessor controlled synchronous motor drive', IEE 3rd International Conf. on Power Electronics and Variable Speed Drives, 13–15 July, London, England, *IEE Conf. Publ. No. 291*, p. 245.

Jahns, T. M. *et al.* (1986) 'Interior permanent-magnet synchronous motors for adjustable-speed drives', *IEEE Trans.*, **IA-22**, No. 4, p. 738.

Jufer, M. and Pofet, A. (1984) 'Synchronous motor with permanent magnets and solid-iron rotor', *ICEM*, Lausanne, p. 598.

Levran, A. and Levi, E. (1984) 'Design of polyphase motors with PM excitation', *IEEE Trans. on Magnetics*, **MAG-20**, No. 3, p. 507, May.

Lloyd, M. R. (1984) 'Survey of electric motors used in variable speed drives', IEE International Conf. on Power Electronics and Variable Speed Drives, 1–4 May, London, England, *IEE Conf. Publ. No. 234*, p. 180.

Morimoto, S. *et al.* (1990) 'Current phase control methods for permanent magnet synchronous motors', *IEEE Trans. on Energy Conf.*, (Power Electronics), Vol. 5, No. 2, p. 133, April.

Parasiliti, F. and Poffet, P. (1989) 'A model for saturation effects in high-field permanent magnet synchronous motors', *IEEE Trans. on Energy Conf.*, Vol. 4, No. 3, p. 487, September.

Platt, D. (1989) 'Permanent magnet synchronous motor with axial flux geometry', *IEEE Trans. on Energy Conf.*, **MAG-25**, No. 4, p. 3076, July.

Rahman, M. A. (1980) 'Permanent magnet synchronous motors—a review of the state of design art', *ICEM*, Athens, Greece, Pt. 1, p. 312.

Rahman, M. A. (1985) 'Analytical model for interior-type permanent magnet synchronous motor', *IEEE Trans.*, **MAG-21**, No. 5, p. 1741, Sept.

Rahman, M. A. and Little, T. A. (1984) 'Dynamic performance analysis of permanent magnet synchronous motors', *IEEE Trans. PAS*, **PAS-103**, No. 6, June.

Rahman, M. A. and Osheiba, A. M. (1990) 'Performance of large line-start permanent magnet synchronous motors', *IEEE Trans. on Energy Conf.*, Vol. 5, No. 1, p. 211, March.

Russenschuck, S. (1990) 'Mathematical Optimisation Techniques for the design of permanent magnet synchronous machines based on numerical field calculation', *IEEE Trans. on Energy Conf.*, **MAG-26**, No. 2, p. 638.

Sebastian, T. *et al.* (1986), 'Modelling of permanent magnet synchronous motors', *IEEE Trans.*, **MAG-22**, No. 5, p. 1069.

Sebastian, T. *et al.* (1986) 'Design considerations for variable speed permanent magnet motors', *ICEM*, Munich, Germany, p. 1099.

Shah, M. R. (1980) 'Practical approach to the inclusion of electromagnetic field non-linearities in the dynamic modelling of large turbogenerators with emphasis on interaction between machine design and system stability aspects', PhD thesis, Virginia Polytechnic Institute and State University, Blacksburg, Virginia, USA.

Stokes, R. W. (1984) 'Recent developments in semiconductor converters for motor drives', IEE Conf. on Power Electronics and Variable Speed Drives, 1–4 May, *IEE Conf. Publ. No. 234*, p.187.

Sneyers, B. *et al.* (1985) 'Field weakening in buried magnet ac motor drives', *IEEE Trans.*, **IA-21**, No. 2, p.398.

7

CAD OF ELECTRICAL MACHINES

Electrical machine design involves the solution of several, complex and interrelated engineering problems. There is no one optimum design and it is fair to say that the design of electrical machines presents a mathematically indeterminate problem. In designing an electric machine we need to consider electromagnetic, thermal and mechanical problems and it takes several iterations to arrive at a commercially viable design which meets all specifications.

The use of digital computers in conjunction with advanced approximation schemes, such as finite difference and finite element, makes it possible to accurately model the machine's behaviour and enables alternate designs to be assessed quickly. Two approaches to electrical machines computer-aided design may be identified as the analysis and the synthesis methods.

In the analysis method, the dimensions, type of construction and details of the materials used are provided by the designer as input data. The computer program is used to calculate the performance characteristics which are then examined by the designer who uses his experience to alter the design towards meeting the specifications. The process is repeated until the design specifications are met.

In the synthesis method, the logical decisions required to modify an initial design are incorporated in the computer program as a number of statements. The initial design itself can be produced by the CAD package provided that the design procedure, for a given class of machines, is included at the computer program front end and that the details of various materials (such as laminations details) are stored in a data library. The main advantage of the synthesis method is that it allows non-resident expertise to be utilised. On the other hand, because of the large number of logical interrelated decisions in the design process, this method is particularly demanding as far as programming effort and skills are concerned.

At the heart of any CAD package there lies at least one modelling scheme. Various equivalent circuits are used successfully to model the electromagnetic and the thermal fields in electric machines. Electromagnetic equivalent circuits for various classes of machines are described in detail in most textbooks dealing with the subject of theory and operation. The equivalent circuit thermal modelling concept is discussed in chapter 2 of this book. Most CAD systems for the numerical analysis of electromagnetic problems are based on the finite element method; the method has proved to be flexible, reliable and effective.

Finite element packages are powerful tools in research and design. By using the computer, it is possible to analyse a number of different geometries and operating conditions, without the need of building a physical prototype. Also, the numerical simulation provides, in most cases, reliable and accurate information about the machine's behaviour irrespective of geometric complexity and material non-linearity.

As most commercially available numerical analysis computer programs are based on the method of finite element, this method and its application to design of electrical machines are discussed in this chapter.

The basic concepts of the finite element technique are presented. By using the equations that describe the variation of potentials within first-order triangular elements, it is shown how the variational approach to finite elements leads to a process of constrained optimisation. One procedure for defining the entries for the global matrix system is described. Practical considerations on problem modelling are included. It is shown how proper planning of a finite element model allows extra flexibility when defining a sequence of related problems. Important considerations on mesh construction, allowing for the simulation of motion, are discussed.

7.1 THE FINITE ELEMENT METHOD

Finite elements were first applied to solve problems of structural analysis and as early as the 1950s the method was applied to problems of heat transfer and fluid flow. In 1970, an article by Silvester and Chari entitled "Finite element solution of saturable magnetic field problems" opened a new era in the field of applied electromagnetism by proposing a formulation capable of dealing with complex geometries and approaching the problem of magnetic non-linearity [7.1]. It was Peter Silvester and his collaborators who provided major contributions to the development of the method and made it so widely applied in electrical engineering today.

In the finite element method, the original field problem domain is divided into a number of sub-domains, or *elements*. The potential distribution within each element is then approximated by a polynomial (*trial function*) and a numerical solution to the field problem is then obtained with respect to some optimal criterion.

Several element shapes are in use, the principal ones being the triangle, the quadrilateral, and curvilinear shapes [7.2]. Element types are defined in terms of the element shape and the order of polynomial interpolation of the *trial function*.

7.1.1 Potentials and energy minimisation

The analysis of electromagnetic field problems is usually facilitated by using potentials rather than field vectors. Poisson's and Laplace's equations can be obtained from Maxwell's equations and solved under specified boundary conditions stated in terms of potentials [7.3]. For example, the partial differential equation describing a magnetostatic problems is:

$$\nabla^2 P = 0 \tag{7.1}$$

where P is the magnetic scalar potential.

The linear stored energy for a magnetostatic problem, W_M, may also be expressed in terms of the magnetic potential as follows:

$$W_M = \frac{\mu}{2}\int_{\text{vol}} (\nabla P)^2 \, dv \tag{7.2}$$

It is seen that the energy is expressed in terms of the squared value of the field gradient, a feature that can be exploited in order to obtain solutions in terms of potentials. This is because the principle of *minimum potential energy* may be applied to many physical systems in a way that minimisation of the system's energy leads to an approximate solution.

The magnetic scalar potential in the above equations is a function of position, i.e. for a two-dimensional analysis in the x–y plane:

$$P = P(x, y) \tag{7.3}$$

In view of this, the energy associated with the potential distribution is a function of a function, i.e. a *functional* [7.4].

7.1.2 The variational method

The implementation of the finite element technique is normally made in conjunction with a mathematical scheme such as the *variational principle* or *the method of weighted residuals*.

When the variational method is applied to solving a partial differential equation describing a field problem, the field equation is not solved directly. Instead, an energy-related functional is defined, expressed as a volume integral over the problem region of an expression posed in terms of the dependent variables of the problem. For a given function P and its partial derivatives P'_x, P'_y and P'_z, a functional F may be defined as follows:

$$F(P) = \int_{\text{vol}} f(P, P'_x, P'_y, P'_z) \, dv \tag{7.4}$$

where:

$$P'_x = \frac{\delta P}{\delta x}; \quad P'_y = \frac{\delta P}{\delta y}; \quad P'_z = \frac{\delta P}{\delta z}. \tag{7.5}$$

The solution process consists of searching for a function P for which the functional F is stationary. In order to obtain a solution by means of this approach, the functional, F, is defined in a way that the correspondent integrand, f, in equation (7.4) satisfies the *Euler equation*:

$$\frac{\delta f}{\delta P} - \frac{\delta}{\delta x}\left(\frac{\delta f}{\delta x'}\right) - \frac{\delta}{\delta y}\left(\frac{\delta f}{\delta y'}\right) - \frac{\delta}{\delta z}\left(\frac{\delta f}{\delta z'}\right) = 0 \tag{7.6}$$

and also, the original problem is subject to *boundary constraints* at the surface, Ω, bounding the problem region. The bounding surface can be taken as the union of two surfaces as follows:

$$\Omega = \Omega_D \bigcup \Omega_N. \tag{7.7}$$

A constant value of the function P is imposed on the surface Ω_D and, therefore, the variation δP is null. This kind of boundary constraint is called *Dirichlet condition.* On the surface, Ω_N, the boundary constraints specify the requirement of zero normal derivative of the function P. This kind of boundary constraint is called *homogeneous Neumann condition.* It is also known as natural boundary conditions because unconstrained functions P which make the functional $F(P)$ stationary about the solution automatically satisfy the requirement of zero normal derivative on Ω_N. For some problems, the constraints are specified either as Dirichlet or homogeneous Neumann; the bounding surface Ω is then taken as $\Omega = \Omega_D$ (Dirichlet problem) or $\Omega = \Omega_N$ (Neumann problem).

Functionals of scalar and vector potentials for electrostatic and magnetostatic problems are well documented in reference [7.5]; the Euler equation corresponding to each potential is derived and the boundary conditions are examined. Reference [7.5] also presents a number of variational formulations for potential problems and gives stationary functionals obtained in terms of field vectors.

7.1.3 First-order triangular elements

In order to apply the finite element technique, it is necessary to subdivide the problem region into subregions called *finite elements.* This subdivision, or *discretisation,* is made in such a way that a single element cannot contain two different media. A group of elements is called *mesh* and, in principle, the size and orientation of the elements is unrestricted. One type of elements, the first-order triangular element, is chosen for the following discussion. This type of element is widely used in electrical engineering. When first-order elements are employed, all nodes must coincide with the vertices of triangles, as illustrated in Figure 7.1.

The first-order triangular element is also known as the triangle with three nodes. Other triangular elements are also used, like the triangle with six nodes or second-order triangular element, containing three nodes coinciding with the vertices and a further three nodes at the middle of the edges. Triangles with three nodes are used in mesh construction and the additional nodes existent in higher-order triangular elements are introduced as a part of the problem definition. This possibility of increasing the number of nodes associated to the elements is seen as a method of extending the order of the local solution variable [7.5]. Finite elements

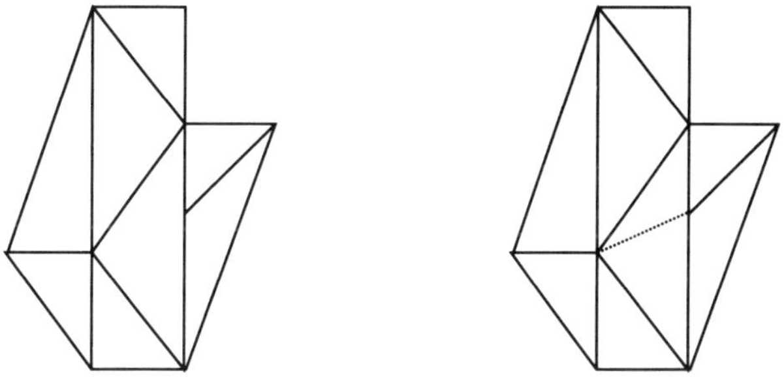

Figure 7.1 Meshes based on first-order triangular elements.

of higher order are sometimes required to obtain more accurate solutions. However, first-order elements, when properly implemented, can provide excellent results and be advantageous with respect to computational cost.

Most finite element formulations are based on potentials. For first-order triangular elements, the potential is assumed to vary linearly within the element as follows:

$$\phi(x, y) = a_1 + a_2 x + a_3 y \tag{7.8}$$

Consider a triangle in the x–y plane, for which the vertices are numbered as 1, 2 and 3. If the potential at these vertices assume values ϕ_1, ϕ_2 and ϕ_3, respectively, equation (7.8) can be applied to each vertex in turn to give:

$$\begin{aligned} \phi_1 &= a_1 + a_2 x_1 + a_3 y_1 \\ \phi_2 &= a_1 + a_2 x_2 + a_3 y_2 \\ \phi_3 &= a_1 + a_2 x_3 + a_3 y_3 \end{aligned} \tag{7.9}$$

The solution of this system of independent simultaneous equations yields the values of the coefficients a_1, a_2 and a_3 in terms of vertex locations and potential values.

In order to obtain an expression for the potential ϕ at a generic location (x, y) within the triangle, the coefficients a_1, a_2 and a_3 are substituted into equation (7.8). After a few algebraic manipulations, the following expression for the local potential is obtained:

$$\phi(x, y) = \frac{1}{D} \sum_{i=1}^{3} (p_i + q_i x + r_i y)\, \phi_i \tag{7.10}$$

where D is equal to twice the area of the triangular element and the geometric coefficients p_i, q_i and r_i are expressed in terms of the coordinates of the element vertices as follows:

$$p_1 = x_2 y_3 - y_2 x_3 \quad q_1 = y_2 - y_3 \quad r_1 = x_3 - x_2 \tag{7.11}$$

$$p_2 = x_3 y_1 - y_3 x_1 \quad q_2 = y_3 - y_1 \quad r_2 = x_1 - x_3 \tag{7.12}$$

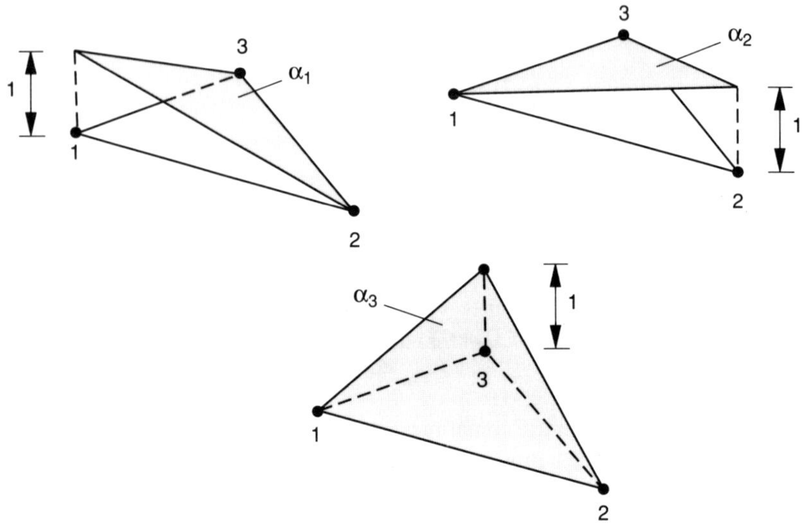

Figure 7.2 An illustration of the planar function αi(*x*, *y*).

$$p_3 = x_1 y_2 - y_1 x_2 \quad q_3 = y_1 - y_2 \quad r_3 = x_2 - x_1 \tag{7.13}$$

The expression $(p_i + q_i x + r_i y)$ in equation (7.10) represents a planar function. By defining:

$$\alpha_i(x, y) = \frac{1}{D}(p_i + q_i x + r_i y) \tag{7.14}$$

equation (7.10) above is reduced to:

$$\phi(x, y) = \sum_{i=1}^{3} \alpha_i(x, y)\phi_i \tag{7.15}$$

Each of the planar functions $\alpha_i(x, y)$ vanishes at two vertices and assumes unity value at the remaining vertex, as shown in Figure 7.2.

Equation (7.15) gives the potential within a single triangular element and this general expression is valid for all the triangles in a given mesh. It should be noted that the potential along any triangle edge is the linear interpolate between its two vertex values. Therefore, the potential will be continuous across the inter-element boundary of triangles sharing two vertices.

7.1.4 Finite element formulation

For a given discretised problem in the x–y plane whose local element potentials are characterised by equation (7.15), the global behaviour of the problem can be established by applying *the variational method.* In this approach, the functional of

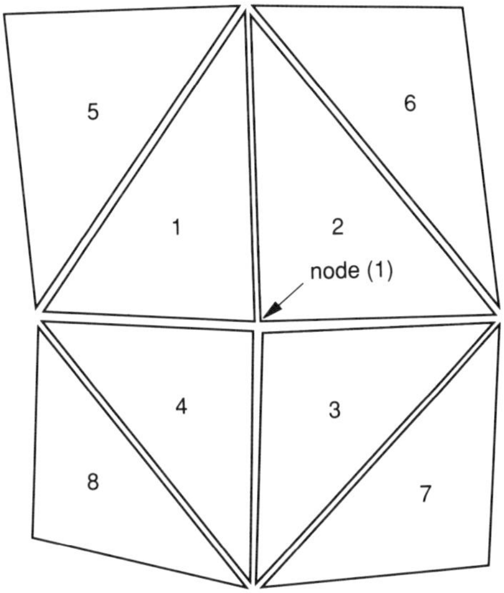

Figure 7.3 Four elements sharing a node.

the problem, F, is approximated as a summation of the individual element energy functional, F_i, as follows:

$$F = \sum_{i=1}^{n_e} F_i \tag{7.16}$$

where n_e is the number of elements in the discretised problem region.

In order to obtain an approximate solution to the problem, the functional is minimised with respect to the unconstrained potentials. Differentiating the functional F given in equation (7.16) with respect to a generic potential ϕ_k and equating to zero we obtain:

$$\frac{\delta F}{\delta \phi_k} = \sum_{i=1}^{n_e} \frac{\delta F_i}{\delta \phi_k} = 0 \tag{7.17}$$

where k ranges over all nodes of unconstrained potentials.

Although the summation of equation (7.17) is extended to all elements in the model, each individual functional F_i is non-zero for the ith element and is zero for all the others. Also, the contribution of a given unconstrained node is due only to the individual functional associated with the elements sharing that node. In the simple case illustrated in Figure 7.3, the contribution to the summation due to the unconstrained node 1 in the centre of the mesh is given by:

$$\frac{\delta F}{\delta \phi_{k|k=1}} = \sum_{j=1}^{4} \frac{\delta F_j}{\delta \phi_1} \tag{7.18}$$

The minimisation of the elements' functional leads to a set of simultaneous equations that can be conveniently expressed in matrix form. Of course, the energy functionals are different for different classes of problems within this vast area of electromagnetic analysis.

The contribution of a single element, say element j, to the summation given in equation (7.17) can be written as follows:

$$\begin{bmatrix} \dfrac{\delta F_j}{\delta \phi_1} \\ \dfrac{\delta F_j}{\delta \phi_2} \\ \dfrac{\delta F_j}{\delta \phi_3} \end{bmatrix} = \begin{bmatrix} S_{11} & S_{12} & S_{13} \\ S_{21} & S_{22} & S_{23} \\ S_{31} & S_{32} & S_{33} \end{bmatrix} \begin{bmatrix} \phi_1 \\ \phi_2 \\ \phi_3 \end{bmatrix} + \begin{bmatrix} Q_1 \\ Q_2 \\ Q_3 \end{bmatrix} \tag{7.19}$$

where the nodes are locally numbered as 1, 2 and 3.

With reference to equation (7.19), the square matrix S_{3x3} that multiplies the array of potentials is known as *stiffness matrix*, a terminology inherited from the structural analysis. Its generic term S_{mn} may be expressed in terms of the element area, its geometric coefficients and κ, a characteristic of the medium, as:

$$S_{mn} = \frac{\kappa}{2D}(q_m q_n + r_m r_n) \qquad m = 1, 2, 3 \quad n = 1, 2, 3 \tag{7.20}$$

The parameter κ represents, for instance, the permittivity for electrostatic problems, and the permeability for magnetostatic problems. The stiffness matrices $S_{3\times3}$ accounting for the contribution of individual elements are, almost, always symmetric. The generic term given by equation (7.20) shows that symmetry is guaranteed if the problem is stationary. Only for a certain class of problems including the movement of some parts of a device is this symmetry lost.

The array $Q_{3\times1}$ represents the sources such as electric charges, electric current and the effect of permanent magnetism.

The global system that accounts for the contributions of all n nodes of the discretised region is expressed by the following matrix equation:

$$SS_{(n\times n)} \cdot \phi_{(n)} = Q_{(n)} \tag{7.21}$$

The matrix $SS_{(n\times n)}$ is the global coefficient matrix. The definition of the entries for this matrix and for the array $Q_{(n)}$ is as follows:

(1) For an element, i, its contribution, expressed as in equation (7.19), is added to the matrix SS and array Q, obeying the global numbering of the nodes;

(2) The Dirichlet constraints are then inserted. If the potential of node m has a prescribed value, ϕ_m, the mth row of the matrix SS will contain a unit value in the diagonal position mm and zeros in all of its remaining positions. Also, the mth position of the array Q contains the prescribed value ϕ_m. This way of inserting a prescribed potential represents a mathematical identity and the corresponding computational manipulation will, of course, preserve the value of ϕ_m.

In most cases, the global coefficient matrix $SS_{(n \times n)}$ has the properties of being symmetric, sparse and banded. These properties are extremely important, specially when dealing with large problems, as they allow the adoption of compressed matrix storage schemes [7.6].

In the above discussion, no assumption is made with respect to the linearity of the parameter κ in equation (7.20). In most cases, the permittivity ε and electrical conductivity σ are constant, but the permeability μ of ferromagnetic materials is characterised by the magnetisation curve which is, almost always, non-linear. In most cases involving electromagnetic field problems it becomes necessary to use values of μ corresponding to the local level of magnetisation, for a group of elements, when assembling the matrix $SS_{(n \times n)}$. In such a case, these values are different for each operating point and it is necessary to use an iterative process. The method of *successive approximations*, which is based on simple linear techniques, can be used for this purpose [7.7].

The remaining task is to solve the linear set of equations (7.21) to obtain the nodal potentials. Many different methods may be used, these include direct and iterative methods. There is an extensive literature on the subject [7.8, 7.9].

7.1.5 Review of commercial finite element packages

The use of the finite element technique for numerically solving the electromagnetic field equations is, nowadays, accomplished by fast, powerful and general-purpose software. There is, in fact, a great interest in the development of general systems, instead of problem-oriented programs.

These systems are known as *electromagnetic CAD systems*, a terminology that stresses the importance of these systems in design. Electromagnetic field simulators or simply finite element packages are other popular terms for describing these systems.

The three main factors that contributed to the development of these systems are: the advent of the digital computer, the development of interactive computer graphics and advances in the area of numerical analysis which allow both geometries and electromagnetic equations to be represented in a discretised form.

Modern software environments that can be classified as electromagnetic CAD systems may be regarded as having three identifiable modules as follows:

(1) A *pre-processor* which enables the user to define the problem's geometry and specify boundary conditions and excitation sources. Mesh generation is conceptually a pre-processing operation and, at this stage, the number of subdivisions of primitives and tolerances are set and materials' properties are specified;

(2) A *solver* for constructing the system of algebraic equations, which models the physical situation mathematically, and produces solutions. Solutions are usually given as a set of nodal potentials;

(3) A *post-processor* that allows the retrieval and analysis of results. At this stage,

it is possible to visualise and manipulate the fields and to calculate important design parameters such as forces, inductances and losses.

There are nowadays numerous electromagnetic CAD systems, some of them commercially available, and these systems may differ significantly in their structures and capabilities. An excellent discussion of the practical issues on finite element analysis systems is given in reference [7.10].

Modern CAD systems incorporate, in the pre- and post-processors, both interactive and batch operations. The solver may be regarded as the principal part of the system, but its operation usually does not require the user's interaction. On the other hand, the modelling of the problem normally involves a number of issues related to the geometry of the device under consideration and the expected characteristics of its fields. The planning of the model may be regarded as a relevant step even for the choice of the appropriate solver, its sequence of solutions and corresponding convergence criteria.

7.1.6 Considerations on problem modelling

When dealing with problems that involve complex geometries, it is very important to plan the finite element model, specially in the aspects concerned with mesh construction and specification of the boundary conditions. This allows extra flexibility for altering material properties, defining sequences of similar problems and also facilitates the analysis of the results.

Some of the practical issues that need to be considered during the planning of finite element models are:

- percentage of the entire structure to be modelled;
- simulation of rotation;
- refinement of discretisation;
- air-gap identification;
- type of finite elements to be used;
- expected level of saturation.

These issues are now discussed in connection with modelling of a brushless d.c. motor with surface mounted, radially oriented permanent magnets. The geometrical details of a motor considered are given in Figure 7.4.

Percentage of structure to be modelled

Sometimes the analysis of the entire structure is computationally expensive and a great deal of time can be saved by analysing only a part of the structure. The minimum section that needs to be modelled is that for which correct boundary conditions can be specified. For electrical machines, the periodic nature of the magnetic field can be utilised and, generally, only one pole pitch needs to be

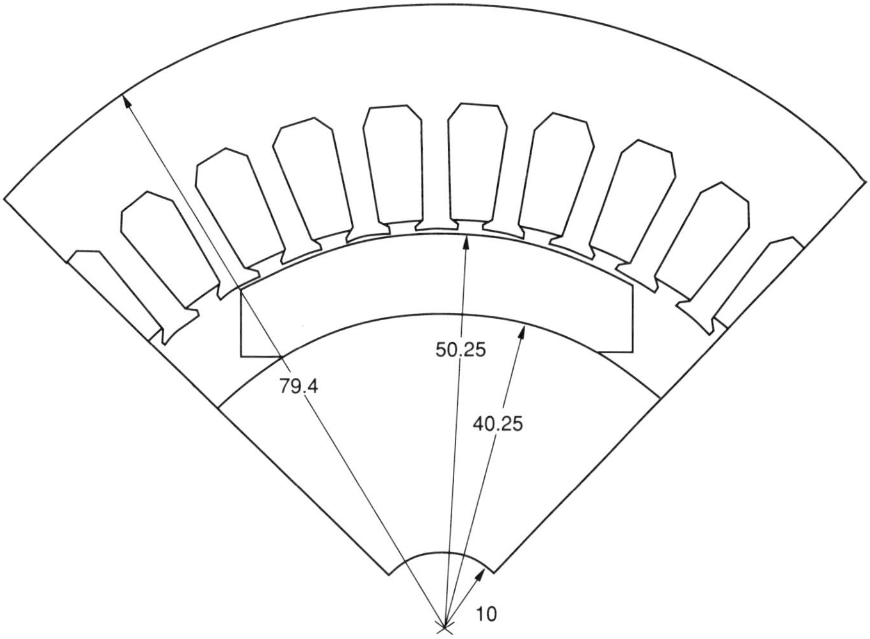

Figure 7.4 Details of a permanent magnet motor.

modelled. The boundary conditions are readily known; these usually consist of, first, enforcing a flux-line boundary along the shaft edge and along the outer periphery of the machine and, secondly, specifying periodicity conditions for the two lateral radial edges separated by one pole pitch.

It is also convenient to place the model in a way that will be symmetrical with respect to one of the coordinate axes. This facilitates not only the input of geometrical coordinates but also the specification of the direction of remanence of existing permanent magnets.

Simulation of rotation

The slip plane is an arc placed in the mean gap and it contains nodes separated at equal intervals [7.11]. This plane is normally used to ease the union of separate models and to allow the implementation of correct mesh deformation, resulting from rotating the inner member, without causing mismatched nodes in the air-gap region.

When modelling electrical machines, it is possible to build two main models, one for the rotor, another for the stator and the slip plane is then taken as an artificial interface between the two models. Each of the two models should include a fraction of the air-gap region, to avoid the construction of a separate model only for this region. For example, the model for the stator, may include the portion of the air-gap region situated between the slip plane and the inner radius of the stator teeth.

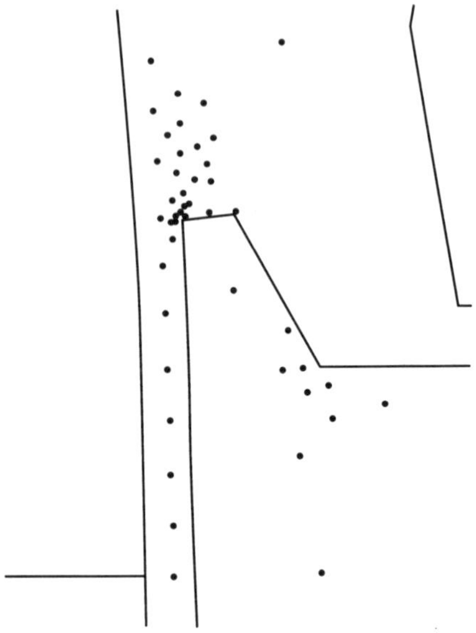

Figure 7.5 Zoomed view of discretisation.

Air-gap discretisation

The mesh around the corners of the stator teeth needs an extra degree of fineness, in order to minimise the effect of the numerical singularity existent at these corners. For this, not only geometrical aspects have to be considered, but also the expected level of saturation; field inhomogeneity is more pronounced in this critical region even at low to moderate levels of saturation.

Figure 7.5 shows a zoomed view of a small region around a stator tooth. In the mean gap the nodes are equally spaced and around the tooth tip the node density is increased.

The radial length of air-gaps in electrical machines is very small and this causes additional difficulties. Ideally, the orientation of the elements in the gap region, and in front of the stator teeth, should follow the way the flux lines are expected to cross the gap, i.e. radially. This clearly, restricts the number of element layers the gap can accommodate. Layers containing regularly shaped elements are necessary for defining integration contours to evaluate forces and torques by the Maxwell stress tensor method, as described later in this chapter.

The stator model

The geometric periodicity of the stator teeth may be exploited in the way that the meshing effort of these critical regions is concentrated on the proper discretisation of an elementary model. This can then be replicated to construct the full model.

An elementary model that spans half a slot pitch is adequate for the manipulation of the mesh. The model is bound, in the circumferential direction, by the slip plane and the motor periphery, and in the radial direction by two radial lines, one coinciding with the centre of a tooth, the other coinciding with the centre of a slot.

In order to choose the separation between the nodes at the slip plane, a simple hand calculation is helpful. This can be done by considering one or two elements accommodated between the slip plane and the inner radius of the stator. The length of the elements in the circumferential direction have to be kept small, to avoid skinny elements. For small air-gaps, an increment of up to 10% of the slot pitch provides a reasonable separation. When simulating the rotor movement, the angle of each rotational step must be equal to, or an integer multiple of, this separation. This is to ensure that nodes on the slip plane always coincide.

The drawing of an outline that represents the upper or lower half of a tooth, part of the back iron and one half of a slot is carried out by entering a few key points and defining the necessary primitives consisting of straight lines and arcs. Next, the need to increase the node density around the tooth corners is considered. This can be done by further subdividing the primitives that intersect at the tooth tip. A first mesh for the model can then be generated and examined. Further alterations of the mesh can be carried out, by increasing the number of subdivisions of the faces or by inserting new nodes and defining new elements. Before moving to the stage where the whole stator is modelled, it is very important to check that the discretisation of the elementary model is satisfactory and that the number of nodes and elements already defined is not too large. If, for example, second-order elements are going to be used in future analyses, the number of nodes will be almost quadruplicated.

Model replication

A model for the stator can be created by replicating the elementary model. The sequence consists of coordinate transformations (such as mirror reflections and rotations) followed by commands for joining the intermediate models.

Sometimes it is possible to make use of true replications, in the sense that the replicated parts retain the same material properties and are not considered new cells or new areas. The use of replications represents a very economical and efficient strategy to build the entire mesh from a small number of regions. However, when this facility is used in problems involving boundary constraints, the areas delimited by the lateral boundaries cannot be modelled by replications. This is because the faces of replicated cells cannot be part of the specified binary constraints. In the present example, replications may be used for modelling most of the area covered by iron laminations and slot openings. The areas accommodating the conductors in the slots are better represented by distinguished regions to allow the specification of the corresponding currents.

The rotor model

The model for the rotor can be created by using the same approach as that used for the stator model, i.e. creation of an elementary model that, after a few geometric transformations results in the full rotor model. The elementary cell for the model may have the shape of a slice, bound circumferentially by the radius of the rotor shaft and that of the slip plane. The angle of separation between the lateral edges can be chosen according to the geometrical arrangement of the individual sectors

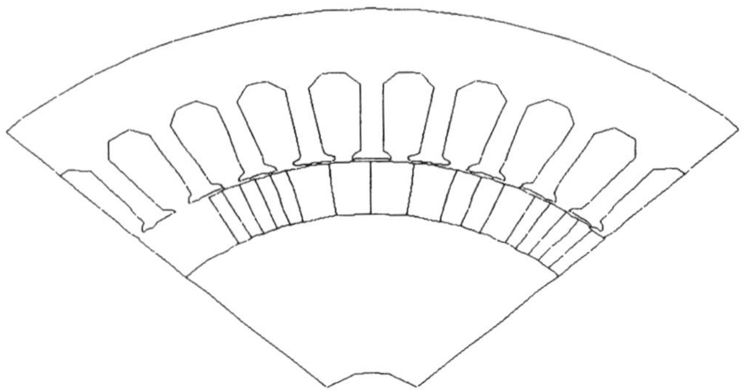

Figure 7.6 Finite element geometric model of a permanent magnet machine.

that represent the permanent magnet. For radially oriented permanent magnets it is necessary to define a number of adjacent sectors, all distinguished by a different label or region number, in order to be able to specify their directions of remanence.

Again, most of the meshing effort is concentrated on the portion of the air-gap. The nodes at the outer periphery of the model should match exactly those of the slip plane.

Once the model of the elementary cell is completed, geometrical transformations can be carried out and, with this, most of the rotor area will be modelled. When the magnet contains a complicated geometry at its edges, further mesh manipulation may be necessary.

The union of models is normally carried out without problems, but it is advisable to check the outline of the resulting model to ensure there are no mismatches.

Simulation of rotation

There are basically two techniques for simulating the rotor movement, namely material redefinition and mesh deformation.

With reference to Figure 7.4, rotation a certain number of degrees in, say, the clockwise direction is simulated by assigning the properties of the magnet material to the air interpolar region next to the leading magnet edge and the properties of air to a section of the same width at the magnet trailing edge. Material redefinition has the advantage of providing a number of solutions on a single mesh. The discretisation errors of the different solutions are more likely to be of similar magnitude and, therefore, this technique is to be preferred when calculating, for example, variation of energy intended for the evaluation of forces or mutual inductances. The model must be planned and constructed to allow utilisation of the concept of material redefinition. For example, the magnet trailing edge region and the interpolar air region should be subdivided into a number of sectors as illustrated in Figure 7.6. Rotation is then simulated by assigning air to one section of the magnet and assigning magnet properties to one air section.

Mesh deformation or mesh distortion is normally used when it is difficult to implement material redefinition (for example, when the magnet edges have complex shapes) or when dealing with a simple geometry for which only two different positions are considered. Mesh deformation may be readily implemented by a parametric transformation applied to all the regions that constitute the part to be distorted. However, the areas of the air-gap region, on both sides of the slip plane and close to the lateral boundaries, have to accommodate the right assembly of elements to allow the correct sequence of deformations and the specification of the periodicity conditions for the subsequent distorted models.

In the motor shown in Figure 7.4, the permanent magnet contains geometric complications at its ends, which make the technique of mesh deformation more convenient. The regions that constitute the stator can be rotated in the counter-clockwise direction in order to simulate the relative movement of the rotor in the opposite direction. As a result of this, the directions of remanence of different sectors of the magnet do not have to be re-calculated and re-specified for each individual problem.

Batch files

When a single geometrical model is to be used to define a sequence of related problems, it is advantageous to exploit the software capability to automate both pre- and post-processing operations. This saves time by avoiding repetitive data input and it minimises the likelihood of human errors.

Batch files are extremely useful specially when analysing sequences of similar problems. These occur very often in analysis and design of electromagnetic devices.

The discussion of problem modelling presented above demonstrates how laborious such a process could be. However, it should be noted that a designer does not have to be an expert on finite element modelling. Most of the modelling time is required for laminations and air-gaps, and as most manufacturers design their motors around a specific range of laminations, it is possible to create finite element models for the laminations range and store them in data files which can then be utilised in the design process. Finite element modelling of intricate motor parts could be done by an expert; not necessarily the motor designer, and once created, these models can easily be integrated, by the designer, into the main model. Such an approach will enable a designer to investigate design changes; for example, the effect of using different magnet materials on the motor's performance, very quickly. Obviously, development of new ranges of motors will require a more lengthy evaluation.

It is possible to write a *software shell*, in which pre- and post-processing steps are completely automated, around a commercial finite element package. This approach gives the designer access to the accuracy of the finite element solutions without having to operate the process directly (or without even knowing that the analysis routine at the heart of the design package utilises numerical field solution techniques) [7.12].

7.2 APPLICATION OF THE FINITE ELEMENT METHOD TO ELECTRICAL MACHINES DESIGN

The application of the finite element method to the design of electrical machines enables determination of important design parameters, such as flux linkages, induced

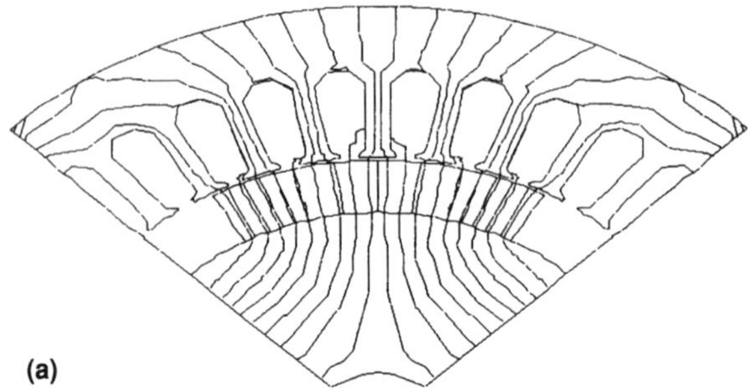
(a)

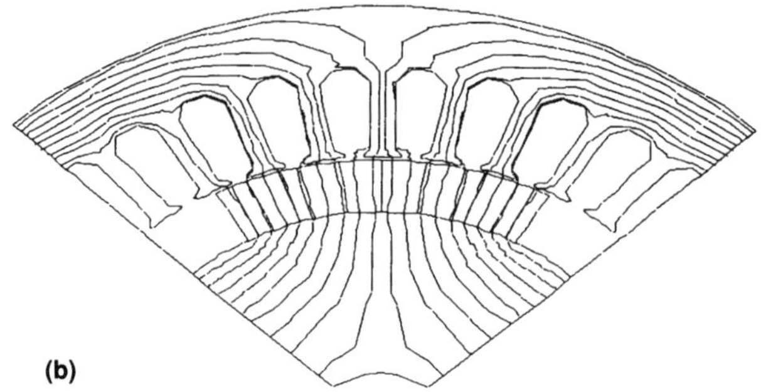
(b)

Figure 7.7 Flux plots showing wrongly (a) and correctly (b) specified boundary conditions.

voltages, core losses, winding inductances and electromagnetic torque developed, with a very high accuracy.

7.2.1 Flux distribution

Prediction of flux distribution is very useful for the design process. It gives the status of magnetic saturation throughout the machine as well as the ratio of leakage to useful flux in various parts of the magnetic circuit. In most cases, the pictorial display of flux distribution is what the designer would examine first. Wrongly assigned material codes and excitation current and improper implementation of boundary conditions can be spotted at once in flux plots. In order to illustrate this, Figure 7.7 shows two flux plots for the brushless d.c. motor whose geometrical details are given in Figure 7.4. In the flux plot of Figure 7.7(a) the flux line boundary at the outer surface of the stator core is not enforced; the homogeneous Neumann boundary condition is assumed and the solution is, therefore, in error. Correct implementation of the boundary conditions results in a flux plot as shown in Figure 7.7(b).

7.2.2 Saturated machine inductance

Winding inductance plays an important role in determining the performance of the machine. For example, in brushless d.c. motors, the rate of change of the motor

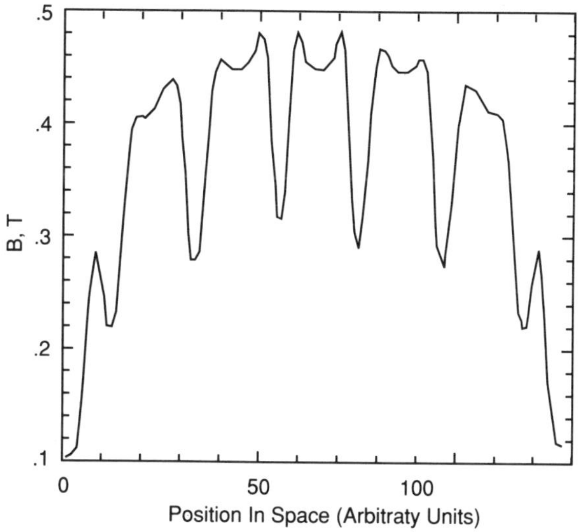

Figure 7.8 Air-gap flux density distribution.

current and its peak and average values, as the d.c. link voltage is switched on and off, are determined by the inductance of the primary winding. In conventional d.c. motors the quality of commutation is determined by the armature inductance.

The effective value of the winding inductance under a specified operating condition may be determined at the design stage from:

$$L = \frac{2W}{i^2} H \tag{7.22}$$

where W is the energy stored in the magnetic field, W, and i is the winding current.

Most commercial finite element packages provide commands for the calculation of energy in all or in sections of the modelled problem.

7.2.3 Flux density and emf waveforms

In most cases, the solution of the electromagnetic field problem is obtained in terms of magnetic vector potentials. The flux density is therefore obtained as the curl of the magnetic vector potential. Most commercial finite element packages have simple commands to invoke the calculation of flux density from the potential solution. The distribution of the flux density in the air-gap can readily be determined by defining a contour along the mid-gap points and retrieving the normal component of the flux density along it. Typical air-gap flux density distribution is shown in Figure 7.8 for a brushless d.c. motor.

The mid-gap radial flux density waveform obtained from the finite element non-linear field solution may be used to calculate the total flux linkage and induced back emf. The variation of flux density in the air-gap over one pole pitch is shown in Figure 7.8 and this waveform can be represented by a Fourier series as follows:

$$B_g(\theta) = \sum_{n=0}^{K_n} [A_n \cos(n\theta) + B_n \sin(n\theta)] \tag{7.23}$$

Where $B_g(\theta)$ is the radial air gap flux density, T, θ is the space angle expressed in electrical degrees, n is the harmonic order, K_n is the total number of harmonics considered and A_n and B_n are the Fourier coefficients.

The harmonic components of the average flux per pole, ϕ_n, are then calculated from the machine geometry and the air-gap flux density as follows:

$$\phi_n = \left(\frac{2}{\pi}\right) L_i \tau_n [A_n \cos(n\theta) + B_n \sin(n\theta)] \tag{7.24}$$

where L_i is the effective machine length, m, and τ_n is the pole pitch for the nth harmonic, m.

Once the harmonic flux per pole is known, the mid-gap emf waveform, E_a for phase a, is obtained from the time rate of change of flux, winding distribution factors (K_{wn}), and the number of series turns per phase (T) as follows:

$$E_a = T \sum_{n=1}^{K_n} -\left(K_{wn} \frac{\delta\phi_n}{\delta t}\right) \tag{7.25}$$

Utilising equation (7.24), and noting that:

$$\theta = \omega t \quad \text{and} \quad \tau_n = \frac{\tau}{n}$$

we obtain:

$$E_a = -\omega \frac{2}{\pi} L_i \tau T \sum_{n=0}^{K_n} [K_{wn}(-A_n \sin(n\omega t) + B_n \cos(n\omega t)] \tag{7.26}$$

The harmonic winding factor (K_{wn}) is calculated in terms of the number of slots per pole per phase (q) and the slot angle (ψ) as follows:

$$K_{wn} = \frac{\sin(nq\psi/2)}{q \sin(n\psi/2)} \tag{7.27}$$

Phase a emf is shifted by 120 and 240 electrical degrees to obtain phase b and c emfs, respectively.

This method of calculating the induced emf yields more accurate results than when just the fundamental component of the air-gap flux density waveform is considered. The number of harmonics that need to be taken into account depends

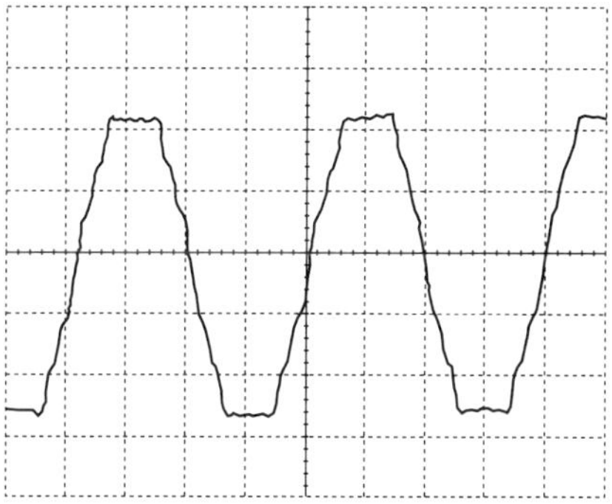

Measured

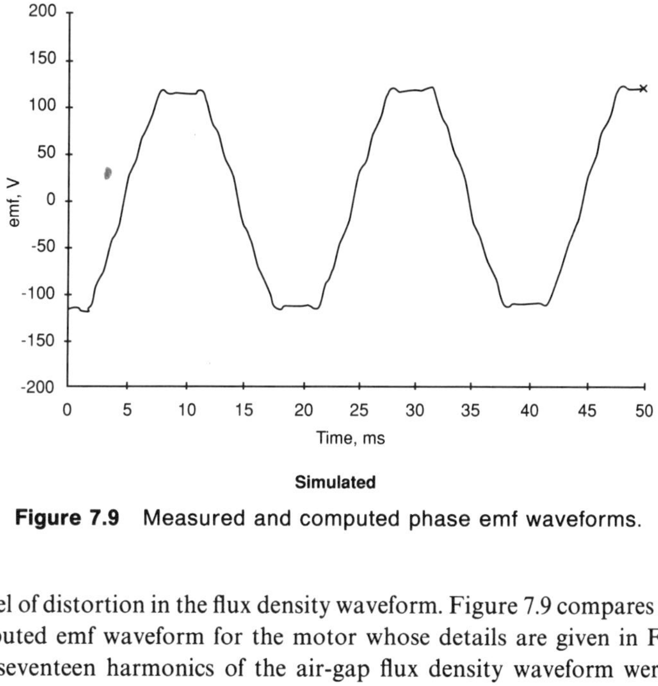

Simulated

Figure 7.9 Measured and computed phase emf waveforms.

on the level of distortion in the flux density waveform. Figure 7.9 compares measured and computed emf waveform for the motor whose details are given in Figure 7.4. The first seventeen harmonics of the air-gap flux density waveform were used to predict the emf waveform. It is seen in Figure 7.9 that the measured and computed waveforms are in close agreement.

7.3 FORCE AND TORQUE CALCULATIONS

The calculation of forces and torques developed in electromagnetic devices, and their variation with changes in position and excitation is, in most cases, what the designer is interested in. The calculated values must be accurate if they are to be of any practical significance.

Most users of finite element packages are able to obtain the field distribution with the level of accuracy required for their purposes. Generally, accurate values of the force developed in electromechanical devices are more difficult to obtain.

The electromagnetic torque developed in an electric machine can be obtained from a numerical field solution by evaluating the Maxwell stress tensor along an arbitrary contour, by the virtual work method, or by body force integration (the *Bil* method). These methods are based on different physical principles and, although their formulations may be found to be equivalent in theory, marked differences are usually found in the numerical results produced by different methods applied to the same problem. The source of numerical errors is different for each method and this is a critical point in the choice of the method. When choosing the method for the evaluation of force and torque it is important to take into account many aspects of the device's geometry and field distribution.

Besides the three popular methods for force and torque evaluation (virtual work, Maxwell stress tensor and *Bil* methods) other methods exist; notably, the mean and difference potential method. The basic formulation and computational implementation of these methods are discussed below. Also, other, less popular methods, are briefly discussed.

7.3.1 The virtual work method

The basis of the method is the relationship between force and energy, provided by the principle of conservation of energy [7.14]. The force acting on a movable part of a device may be evaluated by determining the variation of the stored magnetic co-energy of the entire device when a small displacement takes place. The method is best introduced by considering the simple case where only one-dimensional movements are considered. In such a case, the force can be evaluated by:

$$F = \frac{W_2 - W_1}{x_2 - x_1} \tag{7.28}$$

where W_1 and W_2 represent the stored co-energies at two distinct positions x_1 and x_2 and F is the estimate to the force at the intermediate position.

Equation (7.28) gives the component of the force acting along the direction of movement. When rotational displacements are considered, the corresponding equation to evaluate the torque is:

$$T = \frac{(W_2 - W_1)}{\theta_2 - \theta_1} \tag{7.29}$$

where W_1 and W_2 represent the stored co-energies at two distinct angular positions θ_1 and θ_2 and T is the estimate of the torque at the intermediate position.

The expressions for the force and torque can be expressed more conveniently by the following difference quotients:

$$F = \frac{\delta W}{\delta \theta} \tag{7.30}$$

and

$$T = \frac{\delta W}{\delta \theta} \tag{7.31}$$

The expressions for the force and torque given in equations (7.28) to (7.31) are readily identified as approximate derivatives of a function representing the variation of the device's stored co-energy with respect to positional displacement. Large computational errors may occur in the process of numerical differentiation unless very accurate data are available. This leads to difficulties when implementing the method. These difficulties are mainly related to the choice of the positional displacement and the implications that it has on the accuracy of computed forces and torques.

The numerical problems of the method are mainly related to the accuracy in computing the difference $(W_2 - W_1)$ in equations (7.28) and (7.29). First, there is an error in each of the two energy estimates W_1 and W_2. Even when these two errors considered separately are small, the precision of the difference $(W_1 - W_2)$ depends on the magnitude and sign of both errors. In most cases the energy difference will be even less accurate than the individual energy estimates. The errors in computed energies stem from inaccuracies in potentials, which in turn are affected by the discretisation errors that arise from the finite element model, a subject that is discussed in Section 7.1 above. The cancellation of errors mentioned above is more likely to happen if the discretisation errors are also of similar magnitude and sign. To allow this similarity in discretisation errors to take place, the potential solutions should be computed on very similar meshes or, ideally, on the same one. The accuracy problem is most acute when the energy difference is a very small percentage of the computed energies, because in such a case the effect of the round-off errors will be more severe.

The choice of the positional displacement is a determinant factor in the accuracy problems discussed above. If, in an attempt to improve the accuracy of the approximate derivative, a small displacement is used, the energy values will be of similar magnitude and the subtraction $(W_1 - W_2)$ will be more sensitive to round-off errors. On the other hand, a larger displacement may not be adequate to model the non-linear characteristic which represents the variation of the system's co-energy with respect to position [7.14, 7.15]. If, for example, a curve representing the variation of force or torque against position is to be obtained, important features like the peak force or torque may not be adequately predicted if a large interval is used. Figure 7.10 shows a hypothetical non-linear characteristic representing the variation of co-energy with respect to position. In this illustration, the selected positions d_1 and d_2 quote the range where the most abrupt changes in energy take place (and where the force is maximum) but the slope of the straight line connecting the points W_1 and W_2 is visibly a poor approximation to the derivative and in this case the peak force, for example, will be underestimated.

The prediction of forces and torques is expected to give more accurate results if the implementation is based on double-sided derivatives or differentiation of smooth curves obtained by least-square or min-max procedures [7.14]. In some applications, however, the number of computed stored co-energies is restricted as the result of a large body displacement and the use of refined methods for differentiation will lead to little improvement. The calculation of cogging torque in permanent magnet

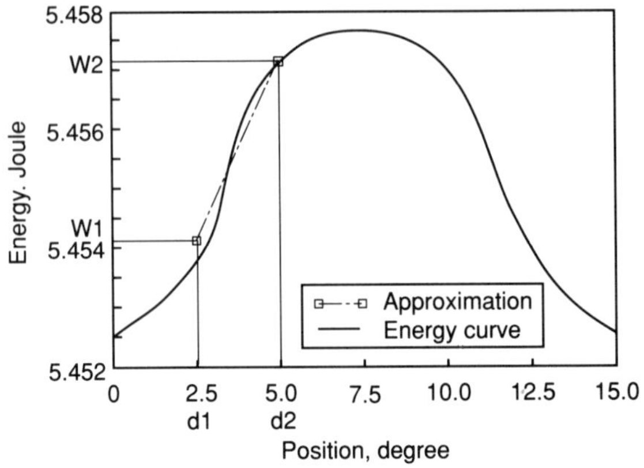

Figure 7.10 A hypothetical variation of co-energy with position.

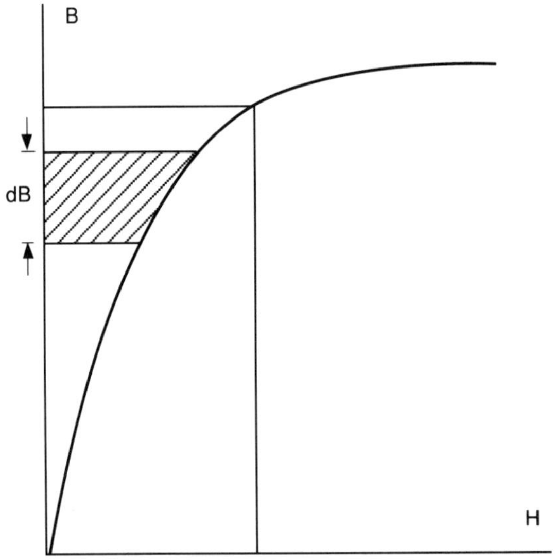

Figure 7.11 Calculation of co-energy.

motors is an example of this sort of application.

The evaluation of the stored co-energy itself is another important point related to the prediction of forces and torques by this method. With reference to Figure 7.11, the stored magnetic co-energy per unit axial length may be obtained by integration over the cross-sectional area of a machine modelled in two dimensions as follows:

$$W = \int_{\Omega} \left[\int_{0}^{B} H \, db \right] d\Omega \tag{7.23}$$

where Ω denotes the modelled problem region.

If the problem is magnetically linear, the stored energy (or co-energy) can also be evaluated by:

$$W = \tfrac{1}{2}\int_{\Omega} AJ\,\mathrm{d}\Omega \qquad (7.33)$$

where J denotes the actual conduction current densities and A is the magnetic vector potential.

While the two formulations are equivalent theoretically [7.16], they differ numerically. The first formulation relies on the derivative of the vector potential and it is, therefore, sensitive to errors in potential values. Since numerical integration is ordinarily an error-attenuating process, results obtained from equation (7.33) are expected to be more accurate [7.17].

Evaluation of stored energy by equation (7.33) over a part of the problem region containing a permanent magnet would require the storage in the data base of the equivalent currents due to the permanent magnet. In cases when only the actual conduction currents associated with current-carrying regions are stored, an alternative way of evaluating the above integral will be required; the equivalent current densities J must be recast explicitly in terms of the coercive force of the magnet material [7.18].

Consider a permanent magnet of length l_0 (in the x–y plane) possessing a straight-line demagnetisation characteristic from $(B_r, 0)$ to $(0, -H_c)$. It can be modelled by surface current sheets $J_s = tJ$ along its edges, comprising a very high current density, J, over a very small thickness, t. By Ampere's law, this leads to a discontinuity of tangential magnetic field $H = J_s$ amperes per metre. For H constant along a magnet edge, and equal to H_c, one can write:

$$\tfrac{1}{2}\int_{\Omega} AJ\,\mathrm{d}\Omega = \tfrac{1}{2}\mathrm{H_c}\int \mathrm{A}\,\mathrm{d}l \qquad (7.34)$$

In this way, integration of the magnetic vector potential along the two sides of the magnet add algebraically (taking into account the sign of the coercivity H_c) to give stored energy [7.18].

With this formulation, it is then possible to avoid numerical differentiation when evaluating the stored energy of any device. This can be done entirely by numerical integration of the magnetic vector potential: the expression given in equation (7.33) may be used for current-carrying regions, and the equivalent line integral expressed in equation (7.34) for permanent magnets [7.18].

The process of evaluating stored co-energy by calculating the area to the left of the B–H curves, expressed in equation (7.32), is more general because the actual non-linear magnetisation characteristics are considered. However, due to the numerical problems that arise from the differentiation of the magnetic vector potential, the process may lead to inaccurate values of energy. There are a number of problems which are, very nearly, magnetically linear and where the magnetic field in the device arises entirely from permanent magnets. Here, it is important to underline two points related to magnetic linearity. First, most permanent-magnet materials used in contemporary technology, ceramics as well as rare earth materials, have linear demagnetisation characteristics. Secondly, there are problems in which the magnetic effects sought represent only a small flux perturbation at any operating point of the device, ensuring incremental linearity.

Were the problem to be treated as magnetically linear, stored energy can be evaluated by integration of the magnetic vector potential, as expressed in equations (7.33) and (7.34). Such an assumption is valid for the problem of predicting cogging torque in unsaturated permanent magnet motors.

7.3.2 The Maxwell stress tensor method

In this method the force or torque is directly calculated from the field distribution. The force and torque are evaluated by integrating the force density over a surface (for three-dimensional models) or a contour (for two-dimensional problems) surrounding the part of interest. The Maxwell stress tensor method is considered to be computationally inexpensive because only one field solution is required to evaluate the force or torque associated with a given rotor position.

Before introducing the mathematical formulation of this method, the relevant terminology and their meaning are given. *Stress* is defined as force per unit area and therefore it can also be referred to as *force density*. Stresses in an electromagnetic field are adequately represented by field *tensors* [7.19]. The expression for the forces transmitted by means of an electromagnetic field may be derived from the Lorentz force expression [7.10].

In electrical machines, it is possible to calculate the total force or torque exerted on a part using only the field conditions on the surrounding surface (or a contour for two dimensional problems) of that part. The magnetic field is taken as a stress transmitting medium in the same sense that a string tying two weights together is the medium that transmits a force from one weight to the other [7.20, 7.21]. A detailed derivation of the expressions for the force and torque is given in reference [7.22]. Those expressions may be adapted for the two-dimensional case where contour integrations, rather than surface integrations, are used.

For a known two-dimensional flux density distribution B and a contour C enclosing a body (or the movable part of a device), the total force and torque acting on the body are given by:

$$F = \int_C \left[\frac{1}{\mu_0} B(B \,.\, n) - \frac{1}{2\mu_0} B^2 \,.\, n \right] \mathrm{d}C \tag{7.35}$$

and

$$T = r \times F \tag{7.36}$$

where r is a vector function with its origin taken at the action point for the torque and n is a unit vector normal to the specified contour.

In the derivation of equation (7.35), two assumptions are made with respect to the integration contour (C):

(1) The contour is entirely placed in air, i.e. it does not intersect magnetic materials or current-carrying elements.

(2) The contour encloses the part on which the force or torque is exerted, but its shape is arbitrarily chosen.

It should be noted that the mathematical concept of total force acting on a rigid body as obtained from the integration of stresses is strictly associated with closed integrals. In some force calculations, however, it is possible to use open contours under special considerations. In a periodic structure, for example, the contour need only embrace one period, the remainder of the structure being taken care of implicitly [7.10]. It is also important to point out that the method can be applied to evaluate forces acting on parts that are not movable; the force of attraction between two stationary magnetic poles separated by an air-gap can be evaluated by integration of the force density along a straight contour situated in the mid-gap. This contour should, for example, extend beyond the poles' lateral dimensions to take into account the effect of fringing.

The integrand of equation (7.35) can be resolved into the components of the force density normal and tangential to the contour, and provide an important tool for the analytical and numerical treatment of force evaluation. These components of the force density are commonly known as *Maxwell's stresses in vacuum* and they are given by:

$$df_n = \frac{1}{2\mu_0}(B_n^2 - B_t^2)\,dC \tag{7.37}$$

and

$$df_t = \frac{1}{\mu_0} B_n B_t\,dC \tag{7.38}$$

There are special cases where, for example, one of the components of flux density is zero and the use of these simpler expressions facilitates the analysis. Consider a known flux density B, constant over a small section dC of a contour parallel to the body surface. Three particular cases are chosen for illustration as follows:

(1) The tangential component of flux density is zero: in this case, $df_t = 0$ and $df_n > 0$; there is only a normal force of attraction.

(2) Both components have the same magnitude: in this case, $df_n = 0$ and df_t is non-zero; there is only a tangential force.

(3) The normal component of flux density is zero: in this case, $df_t = 0$ and $df_n < 0$; there is only a normal repulsion force.

In some applications, only one component of the force is of practical interest. In electrical machines, for instance, the normal force transmitted by the field and accommodated on the bearings is sometimes required. On the other hand, some torque calculations may be based only on the tangential force.

Calculation of torque in electrical machines is made easier if the geometry is modelled in such a way that the action point for the torque coincides with the origin (0, 0) of the cartesian axis. If a circular contour is used, the total tangential force can be obtained by integrating the force density given in equation (7.38); the torque is then calculated using a simple scalar operation, i.e. by multiplying the total tangential force with the radius of the contour.

The computational errors associated with the Maxwell stress tensor method occur, mainly, when obtaining the flux density values from potential solutions by numerical differentiation. Additionally, the location of the integration contour is of crucial importance for the calculation of force and torque.

In the theoretical derivation of the equations for the force and torque, no assumptions are made with respect to the shape of the integration contour, i.e. it can be arbitrarily chosen. In practice, however, the use of different contours in a single field solution yields different values of forces and torques. This is because the numerical field resulting from the finite element solution is, in fact, only an approximation to the true field distribution, i.e. the numerical field contains some sort of error. This error is distributed over the problem region and its distribution is quite often irregular. In other words, the accuracy of the numerical field, generally, differs for different regions of the same problem.

Although there are no strict rules for the choice of the integration contour some recommendations can be made. To start with, contours passing through the nodes must be avoided because flux densities are multi-valued at such nodes. The inconsistencies that might arise from the use of such contours is well illustrated in the work of Mizia *et al.* [7.23]. In their work, torque calculations of a switched reluctance motor using different locations for the integration contour have shown that, for the same solution, results are closer to each other (and to those obtained by the virtual work method) when the contour does not pass through the nodes.

In order to avoid contours passing through the nodes, one or more layers of elements with regular structures should be used in the air-region surrounding the body on which the stresses are to be evaluated. This allows flexibility when defining the location, relative to the position of the nodes, where the contour crosses the elements. Bearing in mind that the contribution of each element to the total force is given by the product of the element's stress component times the length of the path inside the element, it is to be expected that the final result is dependent, to some extent, on the ways the elements are crossed. Ideally, the integration contour should pass through the centres of the elements; for triangular elements [7.24].

Although the evaluation of forces and torques is regarded as a post-processing operation, the above remarks regarding the topological crossing of the elements have to be considered earlier, during mesh construction. Once the potential solution is obtained, the main point of concern is to identify the portions of the solution where the error in the numerical field is comparatively smaller. Different contour locations usually lead to different characteristics of the error magnitude curves as well as to differences in the calculated forces and torques. Results must be examined and compared with measurements, whenever these are available, or with analytical values.

The Maxwell stress tensor method is very sensitive to mesh artefact. Improved local discretisation as well as the symmetry of the mesh have significant influence on the forces and torques computed by this method.

7.3.3 The *Bil* method

The general expression for the force on a current-carrying conductor of length **dl** placed in a region containing a flux density **B** created by other field sources is given by:

$$\mathrm{d}\mathbf{F} = i\,(\mathrm{d}\mathbf{l} \times \mathbf{B}) \tag{7.39}$$

where i is the current.

This expression is normally used for calculating forces between isolated conductors in air or torques in current-carrying loops situated in magnetic fields [7.25, 7.26].

When the flux density and current are at right-angles with respect to each other, it is possible to use a simpler scalar equation to calculate the force developed on a conductor:

$$F = Bil \tag{7.40}$$

Equation (7.40) is considered a convenient way of determining the instantaneous torque in electrical machines. In such an application, it is assumed that the nett current of each slot is confined to an equivalent filamentary conductor of length l. The torque contribution from each slot is determined by calculating the tangential force on the corresponding filamentary conductor as follows:

$$F_t = B_r il \tag{7.41}$$

where B_r is the average value of radial flux density over the slot pitch evaluated on a contour lying in the air-gap [7.27]. This method can be applied to calculate the torque of machines where either the rotor or stator is smooth [7.10, 7.28].

7.3.4 Limitations of the conventional methods

The finite element method produces a solution to the problem in terms of scalar or vector potentials. The solution is only a numerical approximation to the true potential distribution, i.e. there is an inherent error in the potential, commonly described as error in the approximation or shape function [7.5, 7.24]. In both the Maxwell stress tensor and the *Bil* methods, the forces and torques are directly related to magnetic flux densities. Flux densities and magnetic field strengths are obtained from potential solutions by means of numerical differentiation. All the familiar sources of errors are present in numerical differentiation but errors in the approximation function are the most critical, even when small, because they are magnified by differentiation algorithms [7.14]. Therefore, errors in field distributions are, generally, greater than those of the corresponding potential distributions. This helps to explain why some formulations for the prediction of forces and torques are more prone to inaccuracy problems than others.

In the following an attempt is made to enumerate the factors that affect the accuracy of force and torque calculations. To simplify the analysis, it is assumed that the discretisation is appropriate to the problem. This focuses the discussion

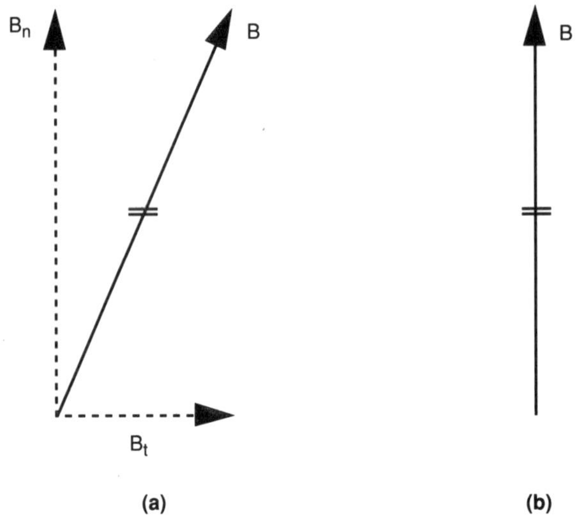

Figure 7.12 Two flux densities of similar magnitude.

on numerical errors that arise from the finite element formulation rather than implementation.

For some algorithms, the quality of the force and torque prediction is dictated by the accuracy that can be achieved in determining the flux distribution. This sensitivity is evident in the Maxwell stress and the *Bil* equations, but not so obviously from the virtual work equations. Here, it is worth remembering that the differences in stored co-energies are due to the different field configurations associated with positional displacement.

Regions where a considerable amount of field co-energy is stored, like geometries with pole tips, are the most critical for the force and torque calculations. In the method of Maxwell stress they contribute the main component of the line integrals in equations (7.37) and (7.38). Similarly, in some formulations of the virtual work method these regions also contribute the main component of the area integral (7.32). In other words, the main contribution to the nett force and torque is due to stresses in areas where pronounced concentration and inhomogeneity of the field occurs; these areas being where an accurate field solution is most difficult to obtain [7.13].

Problems involving the computation of tangential force are often more difficult because the tangential component of the force can be of a much smaller magnitude than the force in the normal direction. This can be illustrated by considering a hypothetical system where the tangential force is small but non-zero. The diagram in Figure 7.12(a) represents the "true" flux density at some point, while the approximate flux density is shown in Figure 7.12(b). Although the magnitude error is small, the incorrect field direction results in a wrong prediction of the tangential force. Therefore, slightly incorrect flux direction is a point of concern and this causes errors in force and torque values calculated by whatever method.

Incorrect flux directions result from the fact that the finite element methods use some form of energy minimisation to drive the convergence. The homogeneous

Neumann boundary conditions, for example, are not exactly satisfied and this affects, even locally, the flux direction [7.5]. Different flux directions imply different values of stored co-energies. Special algorithms are used to perform the numerical differentiation of the shape function and this feature is code-dependent. Two such algorithms are discussed by Binns *et al.* [7.27]. In one of the methods, for example, the continuity of the normal flux density is not imposed at the air–iron interface.

The discussion above identifies the factors which are likely to affect the accuracy of force and torque calculations by any of the three methods. In the following, factors which are unique to each method are identified.

Limitations of the Maxwell stress tensor method

Once the flux density distribution is an approximation to the true one, i.e. there is an inherent error in the numerical field distribution, the independence of the results relative to the choice of the integration contour disappear. The definition of integration contours thus assumes a great importance. This aspect has to be considered very early, during the planning of the finite element model, and adds complexity to the construction of the mesh.

The energy minimisation used by the finite element method produces a numerical solution to the problem that is optimal for a given discretisation. This scheme has no concern for variations in local energy accuracy. Consequently, the resulting fields are globally optimal, even though they may possess considerable local errors. The forces and torques obtained by the Maxwell stress method are related only to flux densities of the elements crossed by a specified contour. This makes it evident why the method is so sensitive to mesh artefact and to the location of the integration contour. Extensive post-processing operations may be necessary to identify the contours that yield reduced error in computed forces and torques. Local error estimation is an elegant alternative for this [7.14].

Limitations of the body force Integration (BII) method

Apart from very special motors designed with air-gap winding or with air-cored magnetic circuits, the vector sum of the forces acting on individual parts of the body is often much smaller than the individual forces themselves. In a conventional motor, the force vectors on individual teeth have very large radial components which, generally, cancel when summed around the air-gap periphery. Errors in calculations increase when the resultant force of interest is calculated in the presence of a much larger force field [7.3].

Limitations of the virtual work method

In contrast to the easy realisation of its formulation, the implementation of the method requires a judicious choice of the positional displacement. If a small system displacement is chosen, the energy values will be of very similar magnitude and this results in significant round-off errors in computing their difference. On the other hand, using a large displacement introduces discretisation errors in the energy–

displacement curve. This choice is problem-dependent and has to take in consideration errors of conflicting nature.

The method requires careful planning of the model and this must be followed by a critical examination of the results. In some cases the results might show that the variation of energy with respect to position is not consistent with the physical realisation of the actual device, and a model redefinition may have to be done. The method is computationally expensive.

The main advantage of the method is the possibility of exploiting the potential distribution directly, as in the evaluation of energy in equations (7.33) and (7.34) for linear problems. Calculations based directly on potential distributions are not part of the usual formulation of the Maxwell stress or the body force integration methods. However, the problem of having two systems with nearly identical energies remains and, in order to overcome this difficulty the computation of the energy difference may be reformulated. This is accomplished by the method of *mean and difference potentials* presented below.

7.3.5 The mean and difference potentials method

Generally, in problems involving both differentiation and integration, as invariably happens in torque prediction, computational errors will be minimised if the order of differentiation and integration can be so rearranged that all integrations are done numerically and all differentiations analytically [7.17]. Since numerical integration is a robust process and analytic differentiation is inherently error-free, the results can be expected to be numerically stable.

The mean and difference potentials method is a new technique for torque computation which have, so far, been applied to permanent magnet machines running at sufficiently low saturation levels [7.29].

The errors associated with the virtual work method stem, mainly, from subtracting two nearly equal energies to obtain a much smaller number representing their difference. The mean and difference potentials method is physically related to the virtual work method but it relies on reformulating the vector potential problem so as to perform subtraction of energies analytically, before problem discretisation [7.29]. Its formulation is described below with reference to the problem of calculating the cogging torque characteristics of a permanent magnet d.c. motor whose constructional details are given in Figure 7.13.

Mathematical formulation

To permit a two-dimensional analysis, the motor is assumed to be infinitely long and longitudinally uniform. All materials are assumed to have constant permeability, and for simplicity, their relative permeability values will be normalised to that of the permanent magnet material used. The stator iron is taken to be sufficiently permeable to confine all flux to the interior of the machine.

Were the virtual work approach to be used, there would be two boundary-value problems to solve, corresponding to, say, rotor positions 1 and 2. These differ only in the placement of the permanent magnetisation. For convenience, the remanence

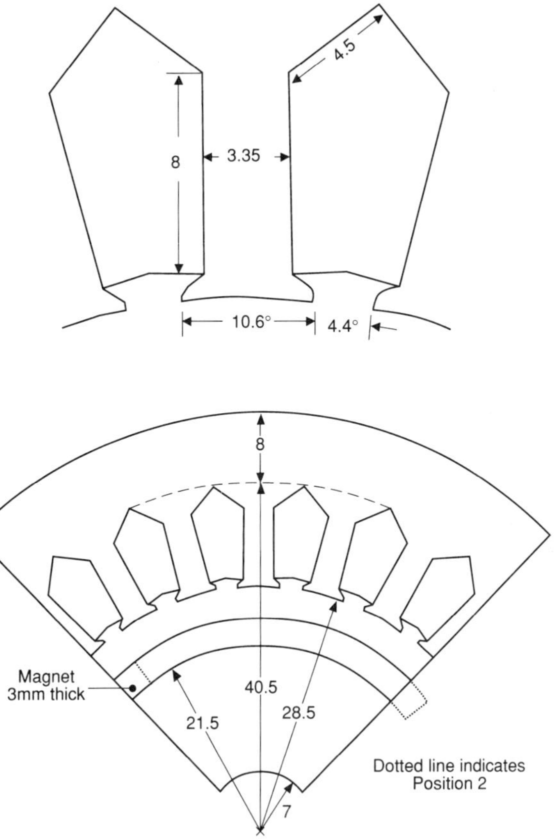

Figure 7.13 Details of a permanent magnet motor.

of the magnets is replaced by equivalent current density distributions J_1 and J_2, corresponding to the two positions. Thus:

$$\mathbf{l}_z \operatorname{curl} \nu \operatorname{curl}(\mathbf{l}_z A_1) = J_1 \qquad \text{in position 1} \tag{7.42}$$

$$\mathbf{l}_z \operatorname{curl} \nu \operatorname{curl}(\mathbf{l}_z A_2) = J_2 \qquad \text{in position 2} \tag{7.43}$$

where $\mathbf{l}_z$ denotes the unit vector in the z direction, along the motor's axial length, A is the magnetic vector potential and ν is the reluctivity.

The boundary conditions that apply to the two problems are similar. The boundary of the problem region Ω may be partitioned into two portions Ω_D and Ω_N, where the vector potential is prescribed on the portion Ω_D, and its derivative on Ω_N, therefore:

$$A_1 = \overline{A} \qquad \text{on } \Omega_D, \tag{7.44}$$

$$\frac{\delta A_1}{\delta n} = \overline{A}' \qquad \text{on } \Omega_N, \tag{7.45}$$

Rather than work with the two boundary-value problems in A_1 and A_2, two new problems are defined. The first is obtained by taking half the difference of the two problems above, the second by taking their mean. Define the *mean potential* by averaging problems 1 and 2,

$$A_m = \frac{A_1 + A_2}{2}, \tag{7.46}$$

with the corresponding current density as:

$$J_m = \frac{J_1 + J_2}{2} \tag{7.47}$$

The mean potential must satisfy:

$$\mathbf{I}_z \operatorname{curl} \nu \operatorname{curl}(\mathbf{I}_z A_m) = J_m, \tag{7.48}$$

subject to the same inhomogeneous boundary conditions (7.44) and (7.45) as the original pair of problems.

Similarly, the *difference* or *deviation potential* is defined by subtracting the two original problems. Therefore:

$$A_d = \frac{A_1 - A_2}{2}, \tag{7.49}$$

with the corresponding current density as:

$$J_d = \frac{J_1 - J_2}{2} \tag{7.50}$$

This potential must satisfy:

$$\mathbf{I}_z \operatorname{curl} \nu \operatorname{curl}(\mathbf{I}_z A_d) = J_d, \tag{7.51}$$

subject, as a result of the subtraction, to the homogeneous boundary conditions:

$$A_d = 0 \qquad \text{on } \Omega_D \tag{7.52}$$

$$\frac{\delta A_d}{\delta n} = 0 \qquad \text{on } \Omega_N \tag{7.53}$$

Were the virtual work method to be used for calculating torque, the stored energy difference δW would be evaluated by computing the energy difference as follows:

$$\delta W = \tfrac{1}{2} \int_\Omega A_1 J_1 \, d\Omega - \tfrac{1}{2} \int_\Omega \mathrm{A}_2 \mathrm{J}_2 \, d\Omega \tag{7.54}$$

This expression involves computing a small difference of large numbers and, therefore, leads to numerical instability. To make it numerically more tractable, it is rewritten in terms of the mean and deviation potentials A_m and A_d. On collecting terms, we get:

$$\delta W = \int_{\Omega} A_d J_m \, d\Omega + \int_{\Omega} A_m J_d \, d\Omega \tag{7.55}$$

Instead of solving the two original boundary value problems, whose solutions are very similar, and then attempting to determine their small difference by subtraction, the energy difference can thus be evaluated by solving the boundary value problems for the mean and deviation potentials. The latter two boundary-value problems are quite different from each other, their solution is nearly orthogonal to each other, and the calculation technique may be expected to be numerically stable.

Implementation and evaluation

To evaluate the method described above, one pole pitch of the same machine shown in Figure 7.13 is modelled, subject to the boundary conditions appropriate to A_d and A_m as expressed in equations (7.44), (7.45), (7.52) and (7.53). The motor is also modelled to enable the cogging torque characteristics to be determined from the conventional virtual work method. In order to provide grounds for comparing the accuracy of the two methods, the same materials characteristics are specified and all potential solutions are obtained with the same mesh. A reasonably fine mesh was used for both methods (952 nodes with 1834 first-order finite elements). Rotor and stator laminations are represented for the purpose of analysis by a linear magnetisation characteristic with a relative permeability of 10 000. The permanent magnet material is represented by a demagnetisation characteristic for which the relative permeability is 1, and for which the coercive force is specified ($H_c = 833$ kA/m, correspondingly $B_r = 1.09$ T).

Rotor displacements were simulated by the technique of material redefinition described in Section 7.2 above. Solutions of two distinct problems are required to enable calculating the energy difference by either method. The period of the cogging torque of the motor shown in Figure 7.13 is 15 mechanical degrees. In this implementation, an incremental rotor movement of 1.5 mechanical degrees (10% of the period) was used, so 20 program runs are required to determine the cogging torque characteristics.

Since this particular motor has no air in the interpolar regions, the field for deviation potential problems is due solely to a sector of magnet spanning 1.5°. For the averaged potential problems, this 1.5° sector is remanence-free but adjacent to this gap there are remanences pointing in opposite directions. This is illustrated in Figure 7.14 where, for simplicity, linearised sketches are used. These, are assumed to span 90 mechanical degrees, the same as the pole pitch. The top two sketches show how the directions of remanence of the magnet change between two consecutive rotor positions, assuming a linear displacement from left to right, i.e. the geometries used for the virtual work method. The other two represent the directions of remanence of the corresponding deviation and averaged problems, respectively.

As the problems are distinguished, basically, by different directions of remanence, pre-processing operations for the corresponding sequence of problems consist, mainly, of a series of commands to edit the material identification labels attached to a group of elements.

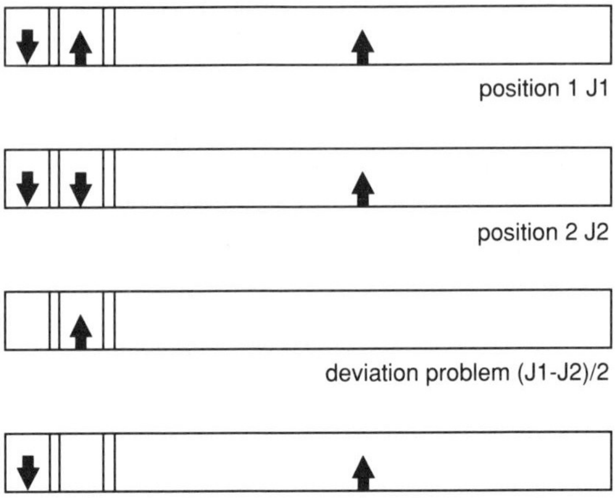

Figure 7.14 Linearised sketches showing direction of magnetisation of different problems.

The use of a single mesh for the different problems in the solution sequence facilitates the combination of solutions when evaluating the energy increment δW expressed in equation (7.55). For any two consecutive rotor positions, an estimate for the torque at the intermediate angle is obtained by:

$$T = \frac{\delta W}{\delta \theta} \tag{7.56}$$

The line integration approach, described in Section 7.3.1, can be used to further simplify the evaluation of the energy increment δW, and this avoids simultaneous retrieval of two distributions, one for the potentials and one for the equivalent current densities. The energy-related surface integrations expressed in equation (7.54) are evaluated, more conveniently, by line integrations. The integral

$$\int_{\Omega} A_{\mathrm{m}} J_{\mathrm{d}} \, \mathrm{d}\Omega \tag{7.57}$$

is replaced by the line integration of the mean potentials, A_{m}, along the edges of the magnet sector where the corresponding current density distribution J_{d} is non-zero. Therefore:

$$\int_{\Omega} A_{\mathrm{m}} J_{\mathrm{d}} \, \mathrm{d}\Omega = \frac{H_{\mathrm{c}}}{2} \left[\int_{C_1} A_{\mathrm{m}} \, \mathrm{d}l - \int_{C_2} A_{\mathrm{m}} \, \mathrm{d}l \right] \tag{7.58}$$

where C_1 and C_2 are two straight contours coinciding with the edges of the magnet sector.

In a similar way, the energy-related integral

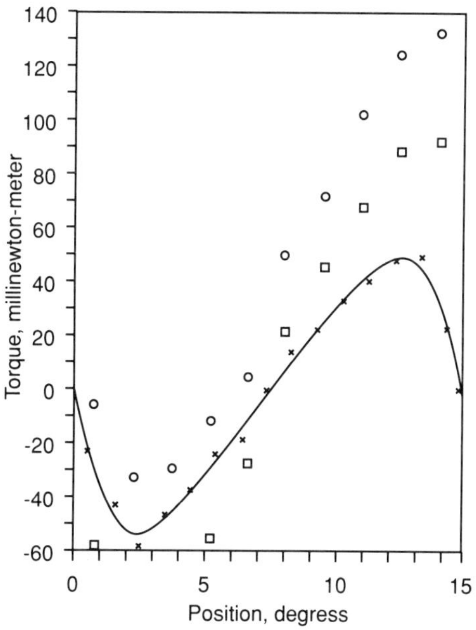

Figure 7.15 Cogging torque computed by the virtual work method. Solid curve represents experimental measurements. Computed results: ○ computed by volume integration, □ by contour integration.

$$\int_{\Omega} A_d J_m \, d\Omega \qquad 7.59)$$

is replaced by the line integration of the deviation potentials, A_d, along the edges of the two sectors of the magnet where the corresponding current density distributions J_m are non-zero.

Numerical results

With the virtual work method, evaluation of energy was performed in two ways; the area integration, equation (7.32), and line integration, equation (7.33). While the two formulations are equivalent theoretically, they differ numerically; there is a difference of about 0.1–0.2% in the computed energies. This difference is overwhelmingly large compared to the energy difference δW between two successive positions, typically about 2–3 millijoules in a total stored energy of about 5.4 joules, i.e. 0.25%. Not surprisingly, the torques computed do not correlate well with those measured. Values of the cogging torques, derived from those energy calculations, are presented in Figure 7.15, along with a curve that represents measured values. Visibly, the stored energy differences affect the accuracy of the computed torques sufficiently to render the straightforward virtual work approach questionable if not completely worthless for this class of problems. Neither curve of Figure 7.15 matches measured results with an acceptable accuracy, and one even fails to show zero average torque over the 15° period. The sensitivity of the virtual work method to numerical error is here seen to be very high.

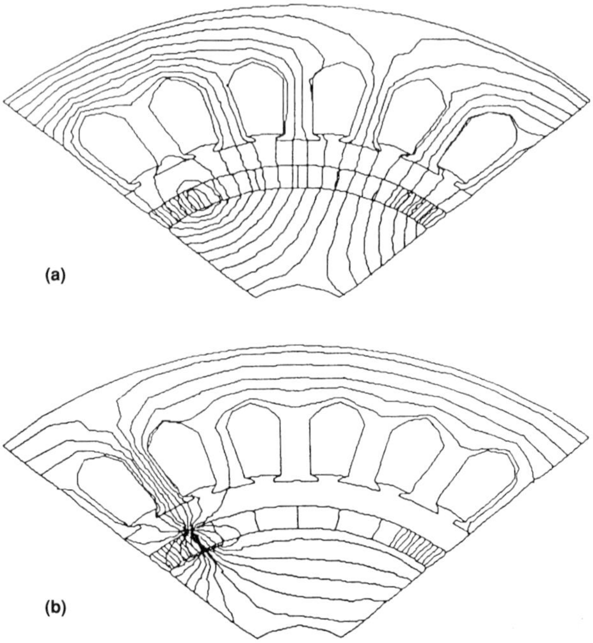

Figure 7.16 Flux plots of the mean and the difference problems.

Two flux plots are used to illustrate the general pattern of the solutions obtained by the mean and difference potentials and these are presented in Figure 7.16. The two problems are defined considering the incremental rotor movement between the angles 13.5° and 15°, and these solutions are used to obtain an estimate for the torque at the intermediate angle, 14.25°.

The flux plots show that the potential distribution of the two problems are quite different from each other. Also, some degree of orthogonality between the two solutions is observed.

The cogging torque characteristic obtained from the mean and deviation potentials method, and computed on the same finite element mesh, is presented in Figure 7.17, together with measurements. The solid curve represents a least-square quintic polynomial fit to the measured values. In contrast to the results of Figure 7.15, which exhibits errors in excess of 100%, the error here is around 3–5%. This error level is considered excellent for a two-dimensional analysis [7.31]. In any case, this percentage error is not very far from the accuracy that can be expected from the experimental results themselves.

In order to investigate how sensitive the mean and difference potential method is to mesh artefact and fineness of discretisation, a second, coarser, mesh was created by deletion of nodes close to the corners of the stator teeth. The number of elements was reduced to 1546, the number of nodes to 808. A comparison of torques obtained from the two meshes by the mean and difference potential method is shown in Figure 7.18, while Figure 7.19 shows mesh details in the air-gap and near tooth tips, where the two meshes differ radically. The two sets of computed torques are

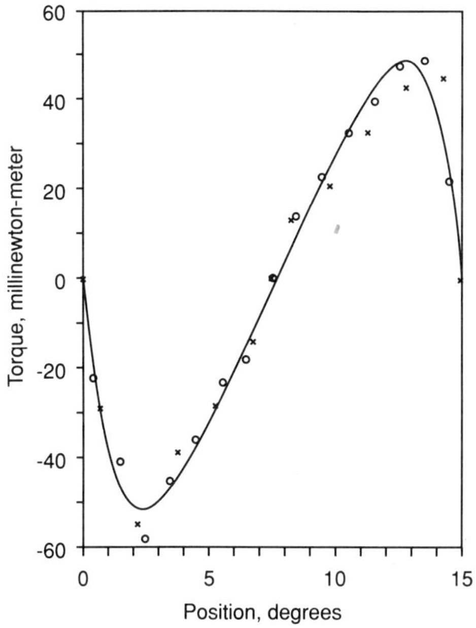

Figure 7.17 Cogging torque computed by the mean and difference potential method. ×, computed points; ○, measured points; solid curve represents quintic polynomial fit to measurements.

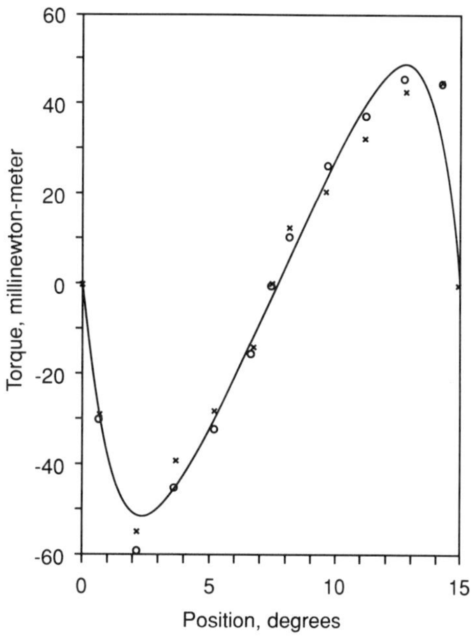

Figure 7.18 Cogging torque computed by the mean and difference potential method, with coarse and fine meshes near the stator tooth-tips. ×, fine mesh; ○, coarse mesh; solid curve represents quintic fit to measurements.

not identical, but both match the measured characteristic equally well. Clearly, the method shows little sensitivity to mesh details.

It should be noted that the flux plots of Figure 7.19 are only superficially similar. While most of the flux lines trace roughly similar courses on the two meshes, tooth-tip flux densities are not similarly distributed so that neither the virtual work method, nor the Maxwell stress calculation, is stable.

7.3.6 Other methods

Rate of change of inductance

Force and torque calculations based on rates of change of inductance were very popular in the past, especially because the formulation incorporates the inductance, an electric circuit parameter which can be measured electrically. The theory for this method is given in many textbooks on electrical machines [7.25, 7.32, 7.33]. The method is closely related to the principle of virtual work and the expressions for the force and torque are:

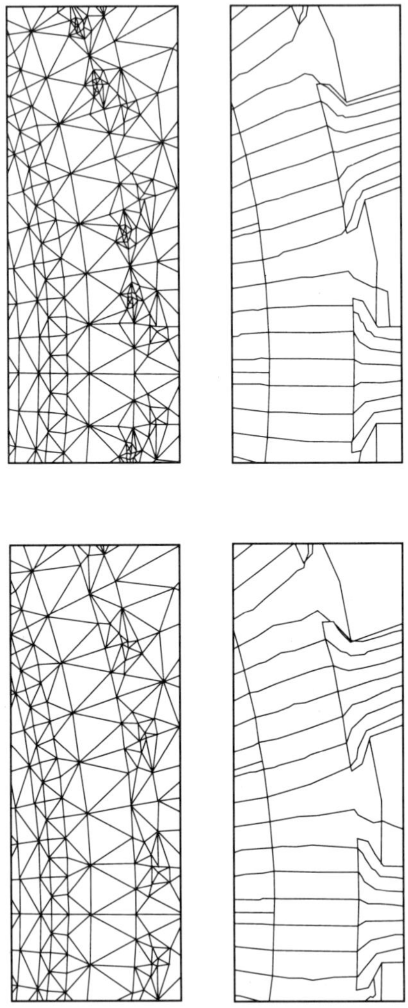

Figure 7.19 Mesh details and flux distributions in the air-gap zone, for two meshes. The upper mesh is finer than the lower, with more than twice as many elements in the air-gap and the tooth-tips.

$$F = \tfrac{1}{2} i^2 \frac{\delta L}{\delta x} \tag{7.60}$$

and

$$T = \tfrac{1}{2} i^2 \frac{\delta L}{\delta \theta} \tag{7.61}$$

The numerical implementation requires two field solutions to determine the variation of inductance using an energy-related approach. The method is computationally expensive and the difference quotients in equations (7.60) and (7.61) are subject to the same kind of computational errors that affect the classic virtual work method.

The single-solution virtual work method

In this method, the magnetised part of the device on which the force or torque acts is taken as a *body* that has been introduced into the system. The work done in introducing the body into the system is taken as the additional stored energy of the system due to the presence of the body [7.19]. The method assumes that as a result of a small displacement, the magnetisation M is kept unchanged and the variation of the system energy is due only to change in the field strength [7.34]. For small displacements δx and δy, the expressions for the x and y components of the force are:

$$F_x = \tfrac{1}{2}\int_{\Omega_b} M_x \frac{\delta B_x}{\delta x}\,\mathrm{d}\Omega_b \tag{7.62}$$

and

$$F_y = \tfrac{1}{2}\int_{\Omega_b} M_y \frac{\delta B_y}{\delta y}\,\mathrm{d}\Omega_b \tag{7.63}$$

where Ω_b denotes the area extending over the magnetised part under consideration.

The torque about the origin can be calculated by

$$T = \tfrac{1}{2}\int_{\Omega_b} M_x \frac{\delta B_x}{\delta x}\frac{y}{r}\,\mathrm{d}\Omega_b + \tfrac{1}{2}\int_{\Omega_b} \mathrm{M}_y \frac{\delta \mathrm{B}_y}{\delta \mathrm{y}}\frac{x}{r}\,\mathrm{d}\Omega_b. \tag{7.64}$$

where $r^2 = x^2 + y^2$.

For the calculation of torque in electrical machines, the integration is performed over the magnetised parts that constitute either the rotor or the stator. Although the whole device has to be meshed, it seems reasonable to choose the frame that contains less saliences for concentrating the meshing effort as well as for performing the integration.

For the evaluation of a single force, it is only necessary to obtain one solution representing the body slightly disturbed from the original position. It is also important to notice that the field values in the air-gap regions are not used in the computations of the above integrals. The method may then be considered as an alternative for force evaluation where a precise modelling of the air-gap regions is difficult to obtain, as in structures containing very small or even infinitesimal air-gaps.

The local jacobian derivative method

Also known as the Coulomb virtual work method, is another formulation related to the principle of virtual work. Only one field solution is needed to compute the virtual energy change. The method is described in reference [7.35] where numerical results are presented for a simple test problem. The nature of the numerical problems associated with this method are discussed in reference [7.14].

7.4 INDUCTION MOTOR ANALYSIS

A difficulty arises when applying the finite element method to induction motors. This is because, in most cases, the motor is voltage-fed and the finite element formulation requires excitation currents to be specified. At any given instant of time, assuming a sinusoidal current variation, the current in each stator slot can be determined in terms of the motor phase current and the winding data. Normal motor operation involves application of a specified voltage to the stator winding and this voltage together with the motor parameters determine the phase currents. The phase current can only be determined if the total motor impedance at a given operating point is known.

Although the design procedure presented in chapter 5 enables determination of the motor's resistances and reactances, the reactance calculation does not take into account the effect of magnetic saturation. The problem is therefore that only the terminal voltage is known and neither the phase current nor the saturated reactance values are known. Iterative techniques are used to overcome this difficulty.

An iterative circuit approach to finite element analysis of induction motors is described in reference [7.36] and its implementation using a commercial finite element analysis package is given in reference [7.37]. The method is illustrated by the flow chart of Figure 7.20. It consists of four identifiable steps as follows, where it is assumed that only one pole pitch is modelled in all finite element problems:

(1) Synchronous speed operation is assumed (i.e. the rotor is stationary with respect to air-gap field) and the magnetising reactance is determined by solving a magnetostatic problem in which the slot currents are defined according to an estimated (or approximately calculated) value of no load phase current and winding details. The stored co-energy is calculated using equation (7.32) and is multiplied with the number of pole pairs and, for two-dimensional analysis, the motor's nett iron length to give the total co-energy, W_t. The magnetising reactance is obtained from:

$$L_m = \frac{2W_t}{mI_2} \tag{7.65}$$

where m is the number of phases and I is the peak value of the assumed no load phase current.

The terminal phase voltage is estimated from:

$$V_{cal} = x_m I = (2\pi f)L_m I \tag{7.66}$$

If the difference between the calculated and the actual terminal voltage is not within an acceptable tolerance, the assumed current will be adjusted by scaling:

$$I_{new} = I\frac{V_{cal}}{V_{true}} \tag{7.67}$$

where I_{new} is the current to be used in the following iteration step.

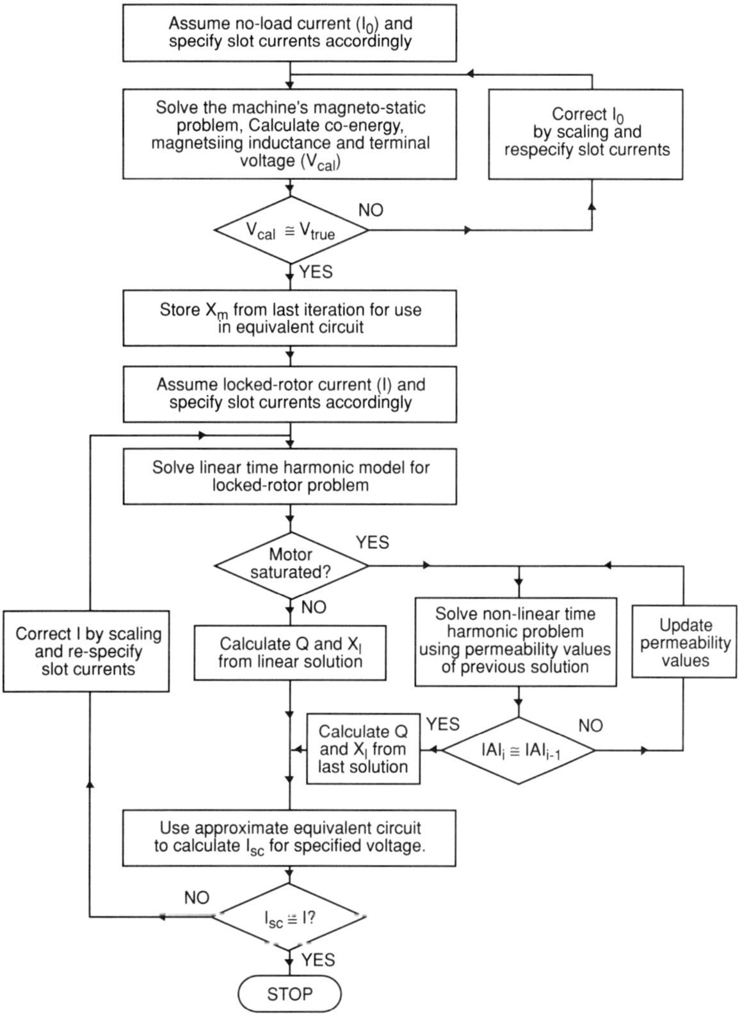

Figure 7.20 Flow chart illustrating the procedure of a circuit approach to finite element analysis of induction motors.

The procedure is repeated until the calculated voltage converges to that applied. The resulting reactance is taken as the saturated value of the magnetising reactance and is used in the motor's equivalent circuit (step 4).

(2) Locked rotor conditions are assumed (i.e. rotor frequency is equal to supply frequency) and the rotor resistance and the combined stator and rotor leakage reactance are determined by solving a time harmonic problem in which the slot currents are defined according to an estimated (or approximately calculated) value of locked-rotor phase current and winding details. In this step a linear problem will be assumed and, if the motor iron is operating in the non linear region of the

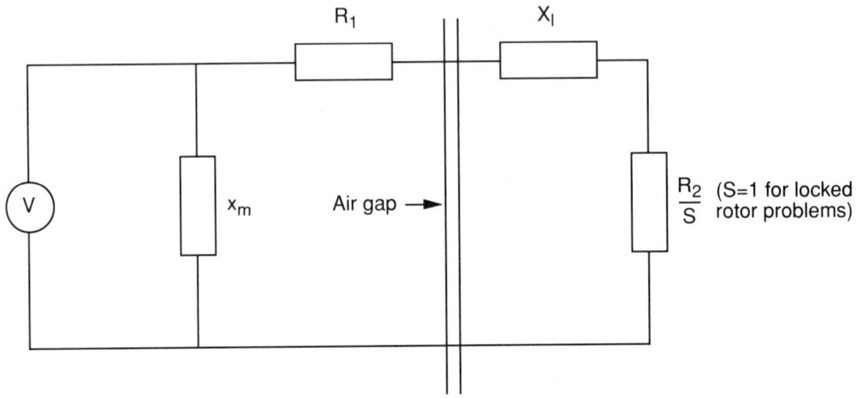

Figure 7.21 An approximate induction motor equivalent circuit.

B–H curve, the reluctivity values will be stored and used in step (3) below, otherwise, the operations of step (4) will be executed.

(3) This is an iterative procedure in which the locked-rotor conditions are assumed. Initially, the reluctivity values resulting from the linear solution of step (2) will be assigned and the non linear problem will be solved. Thereafter, the reluctivity values of the previous solution will be used. The iterative procedure continues until the difference between the infinity norm values of the magnetic vector potential obtained from two successive solutions are within specified limits. When this condition is satisfied, the operations of step (4) will be executed.

(4) The approximate equivalent circuit shown in Figure 7.21 is assumed, where x_1 is the combined stator and rotor leakage reactance. The stator and rotor resistances (r_1 and r_2) are calculated by the design equations given in chapter 5.

The reactive power is calculated from the field solution resulting from steps (2) or (3), for the linear and the non-linear operation, as appropriate [7.37], and multiplied with the nett iron length and number of pole pairs to provide the motor's reactive power, Q. The leakage reactance is calculated from:

$$Q = m(x_m I_o^2 + x_1 I_2^2) \tag{7.68}$$

The phase current corresponding to the applied terminal voltage will be calculated from the equivalent circuit. If the calculated value is widely different from that assumed in the definition of the locked-rotor finite element model, the value used for the model will be modified by scaling and computation is repeated from step (2) above.

The method is reported to yield good results for a number of different motors [7.36]. However, in view of the number of numerical solutions required, this method is computationally very expensive.

7.5 REFERENCES

[7.1] Silvester, P. P. and Chari, M. V. K. (1970) *IEEE Trans.*, Vol. **PAS-89**, No. 7, p. 1642.

[7.2] Zienkiewicz, O. C. (1977) *The Finite Element Method in Engineering Science*, McGraw-Hill, London.

[7.3] Hamdi, E. S. and Licariao-Nogueira, A. F. (1991) *AMSE International Conference on Signals and Systems*, New Delhi, India, AMSE Press, Vol. **3**, p. 111.

[7.4] Riley, K. F. (1974) *Mathematical Methods for the Physical Sciences*, Cambridge University Press.

[7.5] Silvester, P. P. and Ferrari, R. L. (1990) *Finite Elements for Electrical Engineers*, 2nd edn, Cambridge University Press.

[7.6] Hazony, Y. (1984) *Adv. Eng. Software*, Vol. **6**, No. 1, p. 9.

[7.7] Westerberg, A. W. and Berna, T. J. (1978) *Computers and Chemical Engineering*, Vol. **2**, p. 61.

[7.8] Bathe, K. J. and Wilson, E. (1980) *Numerical Methods in Finite Element Analysis*, McGraw-Hill, New York.

[7.9] Irons, B. and Ahmad, S. (1979) *Techniques of Finite Elements*, Ellis Horwood, Chichester.

[7.10] Lowther, D. A. and Silvester, P. P. (1986) *Computer-aided Design in Magnetics*, Springer-Verlag, New York.

[7.11] Preston, T. W., Reece, A. B. J. and Sangha O. S. (1988) *IEEE Trans. on Magnetics*, Vol. **24**, No. 1, p. 471.

[7.12] Ashen, R. A. and Freeman, E. M. (1992) *ICEM 92*, Manchester, England, p. 160.

[7.13] Fano, R. M., Chu, L. J. and Adler, R. B. (1959) *Electromagnetic Fields, Energy and Forces*, John Wiley & Sons, USA.

[7.14] McFee, S. and Lowther, D. A. (1987) *IEEE Trans.*, Vol. **MAG-23**, No. 5, p. 3771.

[7.15] Hamdi, E. S. and Licariao-Nogueira, A. F. (1990) *Advances in Electrical Engineering Software* (P. P. Silvester ed.), Computational Mechanics and Springer-Verlag, Southampton and Heidelberg.

[7.16] Silvester, P. P. (1968) *Modern Electromagnetic Fields*, Prentice-Hall, Englewood Cliffs, New York.

[7.17] Silvester, P. P. (1991) *IEEE Transactions on Magnetics*, Vol. **27**, No. 5, p. 3774.

[7.18] Hamdi, E. S., Isaac, M. L. and Silvester, P .P. (1994) Paper No. P2X16, *CEFC-94*, Aix-les-Bains, France.

[7.19] Stratton, J. A. (1941), *Electromagnetic Theory*, McGraw-Hill, New York and London.

[7.20] Da La Ree, J. and Latorre, J. (1988) *Proc. of IEEE Ind. Appl. Soc. Annual Meeting*, Pittsburgh, USA, p. 32.

[7.21] Panofsky, W. K. and Phillips, M. (1941) *Classical Electricity and Magnetism*, Reading, Mass: Addison-Wesley Publishing Company.

[7.22] Aldefeld, B. (1978) Force in electromagnetic devices, *COMPUMAG-78*, Grenoble, France.

[7.23] Mizia, J., Adamiak, K., Eastham, A. R. and Dawson, G. E. (1988) *IEEE Trans. on Magnetics*, Vol. **24**, No. 1, p. 447.

[7.24] Tarnhuvud, T. and Reichert, K. (1988) *IEEE Transactions on Magnetics*, Vol. **24**, No. 1, p. 443.

[7.25] Gray, C. B. (1989) *Electrical Machines and Drive Systems*, Longman Scientific & Technical, England.

[7.26] Cheng, D. K. (1989) *Field and Wave Electromagnetics*, Reading, Mass: Addison-Wesley Publishing Company.

[7.27] Binns, K. J., Riley, C. P. and Wong, M. (1985) *IEEE Transactions on Magnetics*, Vol. **MAG-21**, No. 6, p. 2435.

[7.28] Binns, K. J. and Lawrenson, P. J. (1973) *Analysis and Computation of Electric and Magnetic Field Problems*, Pergamon Press, Oxford, England.

[7.29] Hamdi, E. S., Licariao-Nogueira, A. F. and Silvester, P. P. (1993) *Proc. IEE* , Pt. A, Vol. **140**, No. 2, p. 151.

[7.30] Licariao-Nogueira, A. F. and Hamdi, E. S. (1991) *AMSE International Conference on Signals and Systems*, New Delhi, India, AMSE Press, Vol. 3, p. 119.

[7.31] Lowther, D. A. and Forghani, B. (1990) *Journal of Applied Physics*, Vol. **67**, No. 9, Pt. 2B, p. 5833.

[7.32] Edwards, J. D. (1991) *Electrical Machines and Drives*, Macmillan Education, Basingstoke and London.

[7.33] Buckley, R. V. (1981) *Electromagnetic Fields*, Macmillan Press, London and Basingstoke.

[7.34] Riley, C. and Simkin, J. (1990) *VECTOR*, Vol. **6**, No. 1, Vector Fields Ltd., Oxford, p. 4.

[7.35] Coulomb, J. L. (1983) *IEEE Transactions on Magnetics*, Vol. **MAG-19**, No. 6, p. 2514.

[7.36] Belmans, R, Findlay, R. D. and Geysen, W. (1990) *IEEE Transactions on Energy Conversion*, Vol. **5**, No. 4, p. 719.

[7.37] Infolytica Corporation (1992) *Procedure for Solving the Induction Machine Problem using MagNet 2D*, Montreal, Canada.

APPENDIX 1

MATERIALS DATA

A1.1 WIRE TABLE (Diameter, mm)

Bare wire	Insulated	Bare wire	Insulated	Bare wire	Insulated
0.040	0.054	0.710	0.790	2.360	2.488
0.045	0.061	0.750	0.832	2.400	2.528
0.050	0.068	0.800	0.885	2.500	2.631
0.056	0.075	0.850	0.937	2.600	2.734
0.063	0.085	0.900	0.990	2.650	2.784
0.071	0.095	0.950	1.041	2.700	2.834
0.080	0.105	1.000	1.093	2.800	2.938
0.090	0.117	1.050	1.143	2.900	3.038
0.100	0.129	1.100	1.196	3.000	3.142
0.112	0.143	1.120	1.217	3.150	3.294
0.125	0.159	1.150	1.247	3.200	3.344
0.140	0.176	1.180	1.279	3.350	3.498
0.160	0.199	1.200	1.300	3.400	3.584
0.180	0.22	1.250	1.351	3.550	3.702
0.200	0.245	1.300	1.403	3.600	3.748
0.224	0.272	1.400	1.506	3.750	3.905
0.240	0.288	1.500	1.608	3.800	3.955
0.250	0.301	1.600	1.711	4.000	4.160
0.270	0.324	1.700	1.813	4.200	4.364
0.280	0.334	1.800	1.916	4.250	4.414
0.315	0.371	1.900	2.018	4.400	4.568
0.355	0.414	2.000	2.120	4.500	4.668
0.400	0.462	2.100	2.223	4.600	4.768
0.450	0.516	2.120	2.243	4.750	4.923
0.500	0.569	2.200	2.326	4.800	4.973
0.560	0.632	2.240	2.366	5.000	5.177
0.630	0.706	2.300	2.428		

A1.2 CONVERSION TABLE – AWG & SWG TO MILLIMETRE SQUARED

Conversion Table AWG – SWG – mm^2

Gauge no.	AWG			SWG		
	Diameter (Inch)	Diameter (mm)	Cross-section (mm^2)	Diameter (Inch)	Diameter (mm)	Cross-section (mm^2)
1	0.2893	7.384	42.41	0.30	7.62	46.50
2	0.2576	6.543	33.62	0.2760	7.01	38.59
3	0.2294	5.827	26.67	0.2520	6.40	32.17
4	0.2043	5.189	21.15	0.2320	5.89	27.25
5	0.1819	4.62	16.77	0.2120	5.38	22.73
6	0.1620	4.115	13.30	0.1920	4.88	18.70
7	0.1443	3.665	10.55	0.1760	4.47	15.69
8	0.1285	3.264	8.367	0.1600	4.06	12.95
9	0.1144	2.906	6.631	0.1440	3.65	10.52
10	0.1019	2.588	5.261	0.1280	3.25	8.30
11	00907	2.30	4.17	0.1160	2.95	6.84
12	0.0808	2.05	3.31	0.1040	2.64	5.47
13	0.0720	1.83	2.63	0.0920	2.34	4.30
14	0.0641	1.63	2.08	0.0800	2.03	3.24
15	0.0571	1.45	1.65	0.0720	1.83	2.63
16	0.0508	1.29	1.31	0.0640	1.63	2.09
17	0.0453	1.15	1.04	0.0560	1.42	1.58
18	0.0403	1.02	0.823	0.0480	1.22	1.17
19	0.0359	0.912	0.635	0.0400	1.02	0.817
20	0.0320	0.813	0.519	0.0360	0.914	0.656
21	0.0285	0.724	0.412	0.0320	0.813	0.519
22	0.0253	0.643	0.324	0.0280	0.711	0.397
23	0.0226	0.574	0.259	0.0240	0.61	0.292
24	0.0201	0.511	0.205	0.0220	0.559	0.245
25	0.0179	0.455	0.162	0.0200	0.508	0.203
26	0.0159	0.404	0.128	0.0108	0.457	0.164
27	0.0142	0.361	0.102	0.0164	0.4166	0.136
28	0.0126	0.320	0.0804	0.0148	0.376	0.111
29	0.0113	0.287	0.0647	0.0136	0.3454	0.0935
30	0.0100	0.254	0.0507	0.0124	0.315	0.0779
31	0.0089	0.226	0.0401	0.0116	0.2946	0.0682
32	0.0080	0.203	0.0324	0.0108	0.2743	0.0592
33	0.0071	0.180	0.0255	0.0100	0.254	0.0507
34	0.0063	0.160	0.0201	0.0092	0.2337	0.0429
35	0.0056	0.142	0.0159	0.0084	0.2134	0.0358
36	0.0050	0.127	0.0127	0.0076	0.193	0.0293
37	0.0045	0.114	0.0103	0.0068	0.1727	0.0234
38	0.0040	0.102	0.00811	0.0060	0.1524	0.0182
39	0.0035	0.069	0.00621	0.0052	0.132	0.0137

Continued

Conversion Table *(Continued)*

Gauge no.	AWG			SWG		
	Diameter		Cross-section	Diameter		Cross-section
	(Inch)	(mm)	(mm^2)	(Inch)	(mm)	(mm^2)
40	0.0031	0.079	0.00487	0.0048	0 1219	0.0117
41	0.0028	0.071	0.00397	0.0044	0.1118	0.0098
42	0.0025	0.064	0.00317	0.0040	0.1016	0.0081
43	0.0022	0.056	0.00245	0.0036	0.0914	0.0065
44	0.0020	0.51	0.00203	0.0032	0.0813	0.0052
45	0.0018	0.047	0.00157	0.0028	0.0711	0.0039
46	0.0016	0.040	0.00125	0.0024	0.0610	0.0029

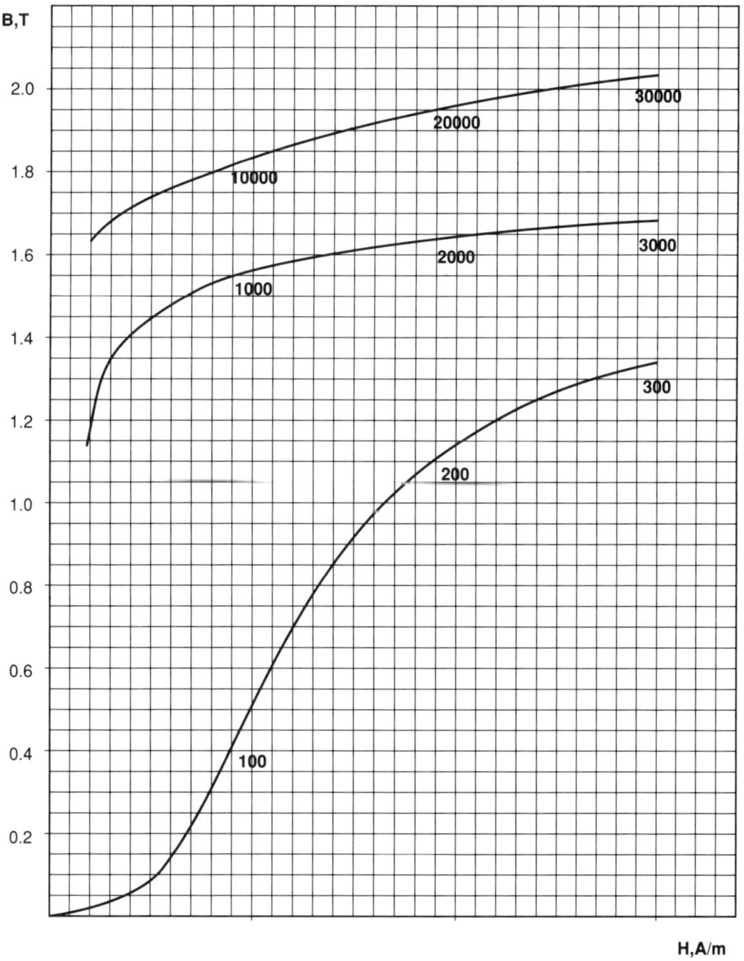

Figure A1.1 Magnetisation characteristics of standard motors laminations.

A1.3 CHARACTERISTICS OF LAMINATION MATERIALS

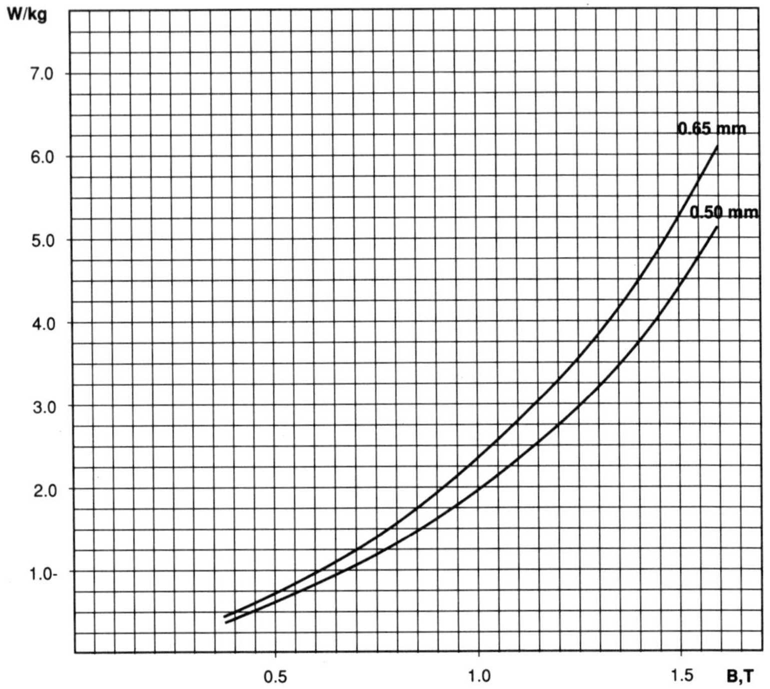

Figure A1.2 Specific core loss at 50 Hz for standard motors laminations.

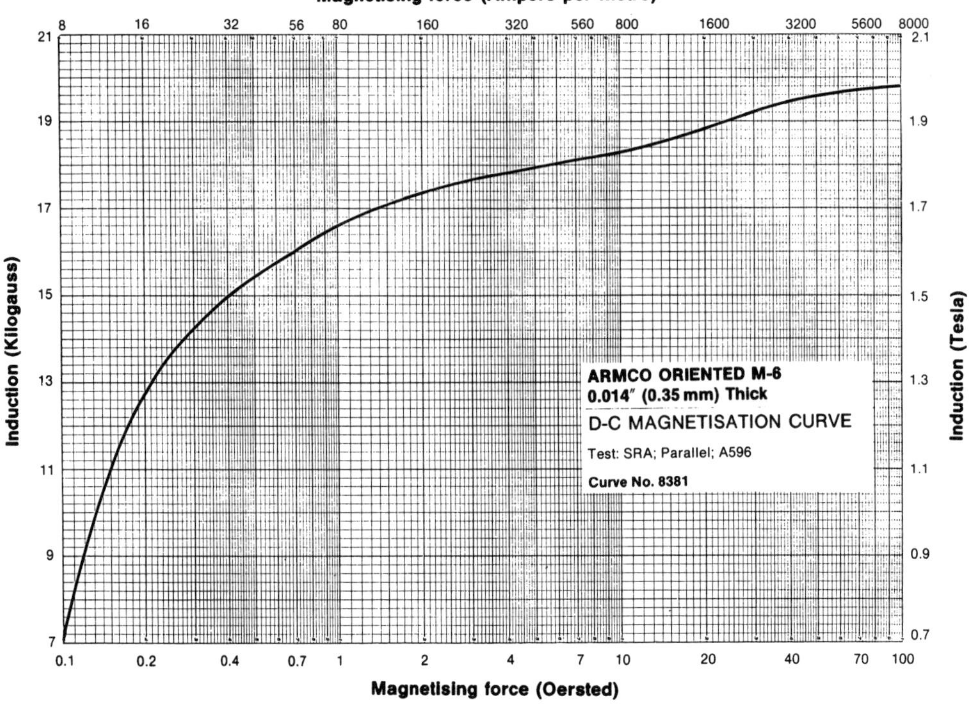

Figure A1.3 Magnetisation characteristics of 400 Hz motors laminations.

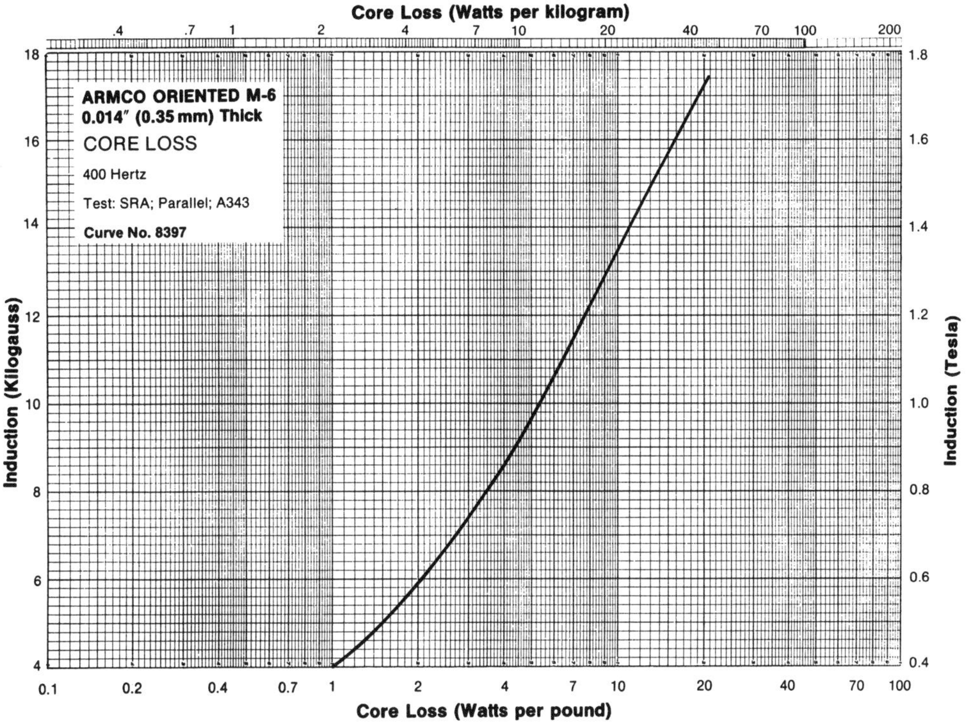

Figure A1.4 Specific core loss for 400 Hz motors laminations.

APPENDIX 2

THE MAGNETIC CIRCUIT OF A d.c. MACHINE

The magnetic circuit of a 4-pole d.c. machine is illustrated in Figure A2.1. The magnetic flux produced by each pole divides, after crossing the air-gap, into two in the armature core. Also, the pole flux is divided into two halves in the stator yoke. Therefore, we may consider the magnetic circuit per pole to consist of the path indicated by the solid line shown in Figure A2.1. The pole ampere-turns required to produce certain flux in the air-gap, ϕ, may be considered to consist of five components: ampere-turns to pass flux in the air-gap (AT_g), to pass flux in the armature teeth (AT_t), to pass flux in the armature core (AT_c), to pass flux in the pole (AT_p) and ampere-turns to pass flux in the stator yoke (AT_y). We will estimate each component separately and these will be added to find the ampere-turns per field pole (AT_{f_0}) required to produce an average flux ϕ in the air-gap at no load.

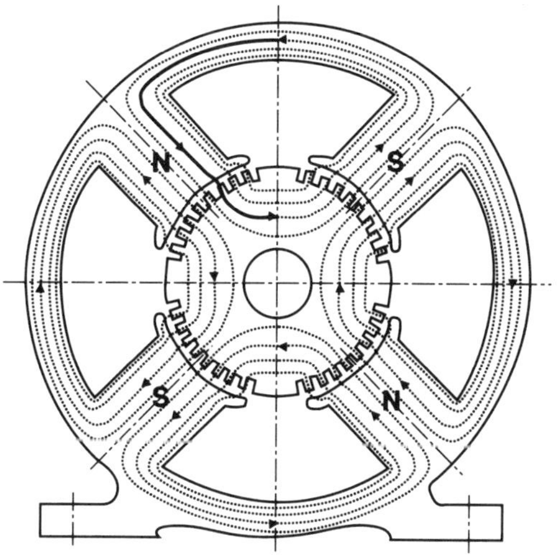

Figure A2.1 The magnetic circuit of a 4-pole d.c. machine.

The effect of the armature field is then discussed and its demagnetising component (AT_{ad}) is calculated. The total ampere-turns per pole required to produce the flux ϕ at full load is:

$$AT_{fl} = AT_{fo} + AT_{ad} \tag{A2.1}$$

A2.1 MAGNETIC CIRCUIT CALCULATIONS

In the following analysis, we will assume that the flux densities in all the iron sections are known and will utilise the relevant magnetisation characteristics of the material to obtain the ampere-turns per metre of the flux path corresponding to the flux density values. The mmf required to send the flux through a particular iron section will then be obtained by multiplying the ampere-turns per metre by the average length of the respective flux path. For the air-gap, the permeability is constant and its ampere-turns vary directly with the flux density (or the voltage). The total mmf per pole at no load and normal voltage is:

$$AT_{fo} = AT_g + AT_t + AT_c + AT_p + AT_y \tag{A2.2}$$

A2.1.1 Air-gap ampere-turns

The ampere-turns required to pass the flux in the air-gap may be readily calculated as the product of the gap's magnetic field intensity (H_g) and its radial length (l_g). The effect of armature slots (slot openings) at the air-gap surface results in the average distance travelled by the flux lines in the air-gap being greater than the actual air-gap length. A correction factor greater than unity, known as Carter's coefficient, is used to account for this. The air-gap ampere-turns are:

$$AT_g = H_g l_g k_g = \frac{B_g}{\mu_0} l_g k_g \tag{A2.3}$$

where B_g is the maximum flux density in the air-gap, T, l_g is the actual air-gap radial length, m and k_g is the air-gap correction factor (or Carter's coefficient).

The air-gap correction factor, which is a function of the ratio of the slot opening to the air-gap length, may be obtained from Figure A2.2.

A2.1.2 Teeth ampere-turns

When calculating the teeth mmf we may consider the flux density in the air-gap to have a constant value equals to B_g over the pole arc, or to have a constant value equals to B_{av} over the pole pitch. In either case, the flux per tooth (or slot) pitch will be:

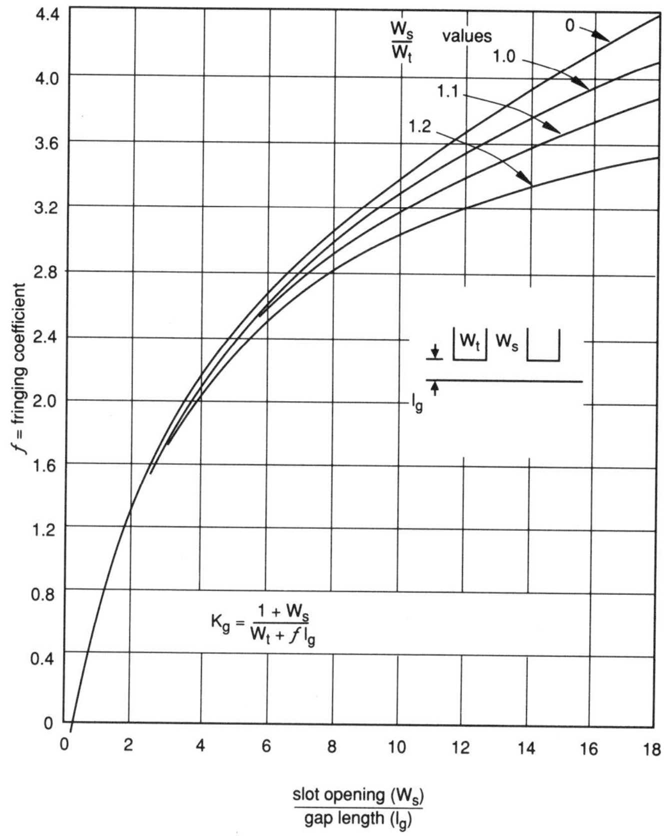

Figure A2.2 The calculation of the air-gap correction factor.

$$\phi_t = \frac{\phi}{\psi(S/p)} \tag{A2.4}$$

where ϕ is the average flux per pole, Wb, ψ is the field form factor, S is the number of armature slots and p is the number of poles.

The apparent flux density in the teeth, B_{ap}, may be obtained as:

$$B_{ap} = \frac{\phi_t}{A_i}, \quad A_i = w_t l_i \tag{A2.5}$$

where w_t is the tooth width, m, and l_i is the effective armature iron length, m.

In the absence of teeth saturation we may assume that all ϕ_t passes through the armature iron, and if parallel-sided teeth are used (as is the case in small motors) we obtain the ampere-turns per metre of the tooth radial length (at_t) from the magnetisation characteristics of lamination material. The teeth ampere-turns will be:

$$AT_t = at_t d_s \tag{A2.6}$$

If the flux density in the teeth is very high saturation will occur. In such a case, not all ϕ_t enters the armature through the iron, some flux lines "leak" through the slots; therefore:

$$\phi_t = \phi_i + \phi_a \tag{A2.7}$$

where ϕ_i is the flux which enters the armature through the tooth iron and ϕ_a is the slot leakage flux.

Dividing both sides of equation (A2.7) by the iron area we get:

$$B_{ap} = \frac{\phi_i}{A_i} + \frac{\phi_a}{A_i} = B_t + B_a\left(\frac{A_a}{A_i}\right) = B_t + B_a K \tag{A2.8}$$

where B_t is the real tooth flux density, T, B_a is the flux density in the slot air, T, A_i and A_a are the iron and air area, respectively and K is the ratio of the air to iron areas.

It should be noted that the slot and teeth form a parallel magnetic circuit and, therefore, the magnetic potential gradient across them is the same. The flux density in the air, (B_a), can be expressed as:

$$B_a = \mu_0 H = \mu_0 at_t \tag{A2.9}$$

where at_t is the mmf per metre corresponding to the real tooth flux density B_t.

The apparent flux density may now be expressed as:

$$B_{ap} = B_t + \mu_0 at_t K = B_t + \mu_0 at_t(K_s - 1) \tag{A2.10}$$

where K_s is the slot coefficient which is defined as follows:

$$K_s = \frac{\text{total slot pitch area}}{\text{iron area}} = \frac{L\tau_s}{l_i w_t} \tag{A2.11}$$

It is now possible to construct the apparent magnetisation characteristics (B_{ap} versus at_t) for various values of slot coefficients. Such characteristics enable determination of the ampere-turns per metre (at_t) corresponding to the real tooth flux density (B_t) when only B_{ap} and K_s are known.

When tapered teeth are used, the tooth flux density will increase, and the permeability of the iron will decrease, as flux lines pass from the top (armature surface) to the bottom of the tooth. In such a case, the ampere-turns per metre corresponding to the flux density one third slot depth from the minimum tooth width is used. The width of the armature tooth at a point one third slot depth from the minimum tooth width is:

$$w_{t3} = \frac{\pi(D - 1.33d_s)}{S} - w_s \tag{A2.12}$$

where d_s and w_s are the slot depth and width respectively, m, D is the armature diameter, m, and S is the number of armature slots.

A2.1.3 Armature core ampere-turns

It is apparent from Figure A2.1 that each section of the armature iron beneath the slots carries half the flux per pole ϕ. If the depth of the armature core below the slots is d_c and the nett armature iron length is l_i, the flux density in the armature core will be:

$$B_c = \frac{\phi/2}{l_i d_c} \tag{A2.13}$$

and

$$d_c = \frac{D - 2d_s - D_i}{2} \tag{A2.14}$$

where D and D_i are the outer and inner armature diameter respectively, m, and d_s is the slot depth, m.

The ampere-turns per metre (at_c) corresponding to the value B_c are obtained from the magnetisation characteristics of armature laminations material. The length of the flux path (l_c) may be taken equal to half the pole pitch on the mean circumference of the armature core:

$$l_c = \frac{\pi(D - 2d_s - d_c)}{2p} \tag{A2.15}$$

The ampere-turns per pole required to pass the flux in the armature core is:

$$AT_c = at_c l_c \tag{A2.16}$$

The flux in the armature core of commutating pole machines consists of two components, one due to the main pole and the other due to the commutating pole. This results in the core flux density increasing on one side and decreasing on the other. In order to avoid saturation, the flux density in the armature core is generally chosen at a value well below saturation; typically 1.5 T.

A2.1.4 Ampere-turns for the pole body

The magnetic flux per pole may be considered to consist of two parts; ϕ, which crosses the air-gap into the armature, and ϕ_l, the pole leakage flux which passes between the poles. The ratio of the flux in the pole body to the air-gap flux is called the pole leakage factor, σ_p. Therefore:

$$\sigma_p = \frac{\phi_p}{\phi} = 1 + \frac{\phi_l}{\phi} \tag{A2.17}$$

The flux in the pole body is not constant for all sections; it has a maximum value near the yoke and it decreases towards the pole shoe. However, to calculate the flux density in the pole we may assume that the flux in the pole is uniform and equal to $\sigma_p\phi$. The cross-section area of the pole is the product of the its width, b_p, and its axial iron length, l_{pi}. The flux density in the pole is:

$$B_p = \frac{\sigma_p\phi}{A_p}, \qquad A_p = l_{pi}b_p \tag{A2.18}$$

The maximum pole flux density, B_p, should not exceed 1.2 T or 1.8 T for solid (cast steel) and laminated poles, respectively.

The ampere-turns per metre, (at_p), is obtained from the magnetisation curve of the pole material. The length of the flux path in the pole body is equal to the pole height, (h_p), plus the thickness of the pole shoe (h_s); therefore:

$$AT_p = at_p(h_p + h_s) \tag{A2.19}$$

A2.1.5 Yoke ampere-turns

The yoke carries half the flux of the pole body. The flux density in the yoke is:

$$B_y = \frac{\phi_p/2}{l_{iy}d_y} \tag{A2.20}$$

and;

$$d_y = \frac{D_{yo} - D_{yi}}{2} \tag{A2.21}$$

where D_{yo} and D_{yi} are the yoke's outer and inner diameter respectively, m, d_y is the yoke depth, m, and l_{iy} is nett iron length of the yoke, m.

The ampere-turns per metre (at_y) corresponding to the value B_y are obtained from the magnetisation characteristics of the yoke's material. The length of the flux path (l_y) may be taken equal to half the pole pitch on the mean circumference of the yoke:

$$l_y = \frac{\pi(D + 2l_g + 2h_s + 2h_p + d_y)}{2p} \tag{A2.22}$$

The ampere-turns per pole required to pass the flux in the yoke is:

$$AT_y = at_yl_y \tag{A2.23}$$

A2.2 ARMATURE REACTION

The axes of the armature mmf lie on the geometric neutral axes and the armature mmf distribution may be approximated to a symmetrical triangular waveform as

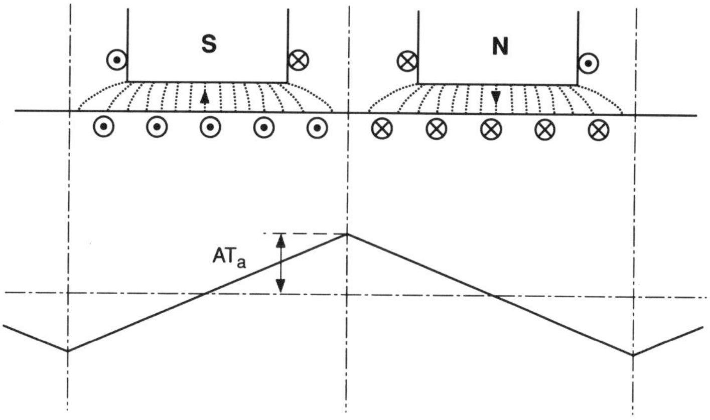

Figure A2.3 Armature reaction mmf.

shown in Figure A2.3. The amplitude of the armature reaction mmf is:

$$AT_a = \frac{I_z Z}{2p}, \qquad I_z = \frac{I_a}{a} \tag{A2.24}$$

where Z is the total number of armature conductors, p is the number of poles, a is the number of parallel armature circuits.

The distribution of the flux density due to the armature mmf under the pole face is similar to the mmf waveform. This is because the air-gap radial length is assumed constant over the pole arc. Although the magnitude of the armature mmf is largest on the interpolar axis, the resulting flux density is much less than that under the pole; because the reluctance of the magnetic circuit increases substantially in the interpolar region.

Figure A2.4 illustrates the pole flux density due to the field mmf, the armature reaction flux density and the resultant air-gap flux density. The resultant flux density distribution is obtained using the principle of superposition; this can only be applied if magnetic saturation is entirely absent. It is shown that the effect of the armature reaction field is to reduce the mmf, and the flux density, under the trailing pole tip and to increase it (by the same amount in the absence of saturation) under the leading pole tip in the case of a motor. The area under the resultant flux density waveform is the same as the area under the flux density distribution curve due to the field mmf at no load. Therefore, in the absence of saturation, the effect of the armature reaction is to redistribute the flux over the pole pitch without affecting the magnitude of the flux produced by the pole excitation at no load. This effect is termed cross magnetisation.

The cross-magnetisation effect of the armature reaction increases the flux density under the one pole tip and this results in magnetic saturation (even if the machine was unsaturated under no load). As a result of saturation occurring at the leading pole tip of a motor (or the trailing pole tip for a generator) the increase in the flux density at the leading pole tip will be less than the flux density reduction at the trailing pole tip; which remains unsaturated. Therefore, the cross-magnetisation

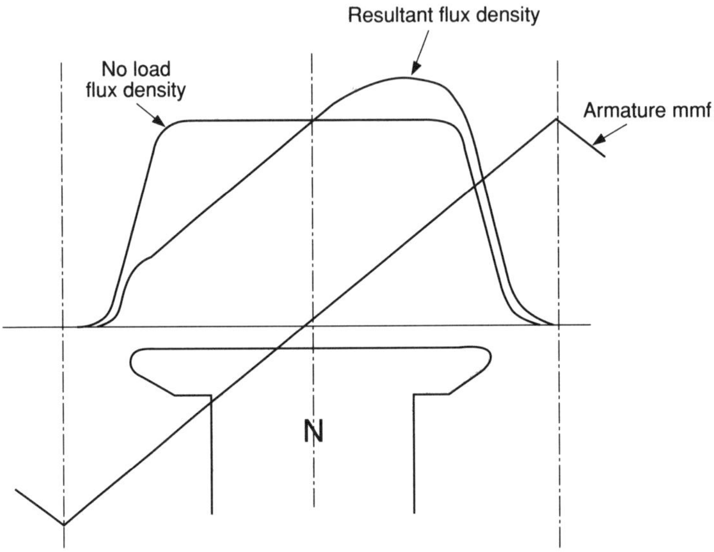

Figure A2.4 Flux density distribution on load and no load.

effect of the armature reaction gives rise to a demagnetising effect. The nett result is that the average flux density under the pole face is reduced and, consequently, the flux per pole is reduced. The cross-magnetisation effect shifts the position of the flux density distribution axis from the geometric interpolar axis, against the direction of rotation for motors. To obtain satisfactory commutation in non-commutating pole machines, the brushes are shifted from the no load interpolar axis against the direction of rotation for motors (and in the direction of rotation for generators); otherwise the flux entering the armature will be in the same direction as the armature reaction mmf and this delays commutation. It should be noted that brush shifting is only effective in unidirectional machines.

Figure A2.5 shows the distribution of the armature current in a 2-pole machine when the brushes are shifted by an angle α. The effect of this brush shift is to divide the armature winding so that a belt of conductors spanning an angle equals to 2α produces a demagnetising field and the rest of the conductors produce a cross-magnetising field.

The demagnetising component of armature reaction ampere-turns acts directly against the field ampere-turns and its magnitude is given by:

$$AT_{\text{ad}} = \frac{2\alpha}{\pi} \frac{I_z Z}{2p} \tag{A2.25}$$

The ampere-turns per pole must be increased above that required to produce the air-gap flux ϕ at no load by an amount equals to AT_{ad} to compensate for the demagnetising component of the armature field.

The cross-magnetising component of the armature reaction produces a magnetic field perpendicular to the pole magnetic field and its magnitude is given by:

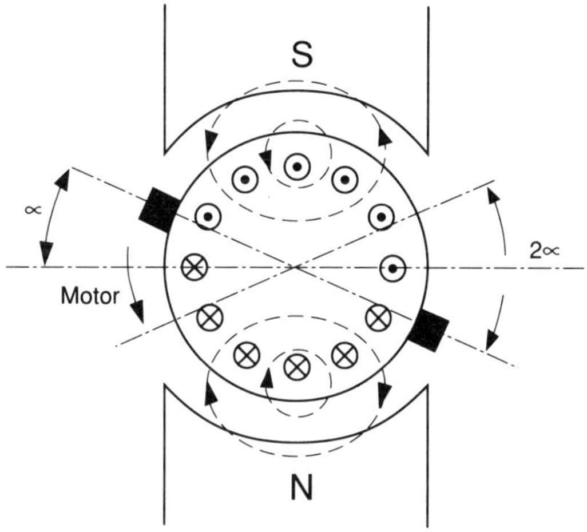

Figure A2.5 Demagnetising and cross-magnetising components of armature reaction.

$$AT_{aq} = \frac{\pi - 2\alpha}{\pi}\frac{I_z Z}{2p} \tag{A2.26}$$

It should be noted that the angle α used in the above equations is the electrical angle of brush shift in radians.

When the machine has commutating poles, the commutating pole produces the commutating field to ensure good commutation, and the brushes remain on the no load neutral position. As a result, all the armature reaction effect is a cross-magnetisation effect in commutating pole machines. However, cross-magnetising armature reaction mmf is likely to saturate the magnetic circuit at one pole tip, as described above, and this gives rise to a demagnetising component of armature reaction. In such cases, the magnitude of the demagnetising component of armature reaction mmf is estimated with the aid of the machine's magnetisation characteristics as described below.

Figure A2.6 shows a typical magnetisation curve which gives the flux produced by the pole in the air-gap (or the generated emf of the machine) as a function of ampere-turns on the pole (or the field current). With reference to Figure A2.6, let us assume that the pole ampere-turns correspond to point P on the horizontal axis. At no load, the flux produced in the air-gap is given by point b on the magnetisation curve. Subtracting the value of armature reaction mmf at the pole tip (ψAT_a) from one side and adding it on the other side of the mmf represented by point P, we get the resultant mmf at both pole tips as points d and f, respectively. The flux at the pole tips is represented by points a and c on the magnetisation curve. As the armature reaction mmf is zero at the centre of the pole and increases, approximately, linearly towards the geometrical neutral axes, we may consider point P to lie at the centre of the pole and consider the curves ba and bc to give the resultant mmf distribution under the pole face. The nett average pole flux is represented by the area under the magnetisation curve from point a to point c divided by the length df; which is less than Pb. This is because the area agb is greater than the area chb.

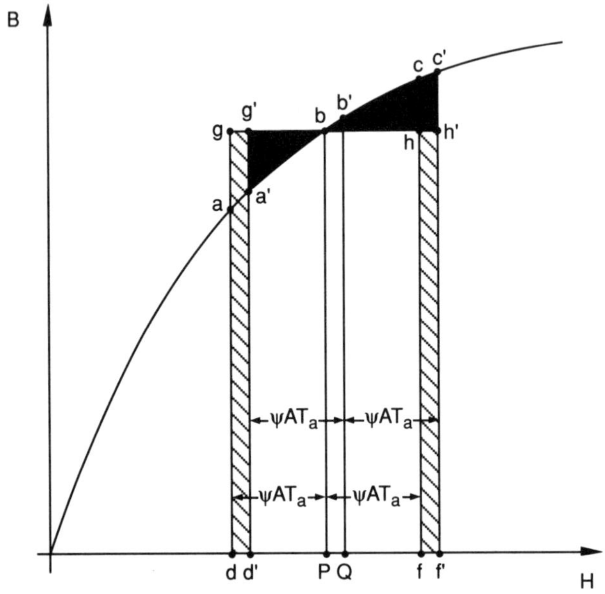

Figure A2.6 Demagnetising component of armature reaction due to magnetic saturation.

In order to obtain an average flux per pole equal to that corresponding to point *b* when the armature reaction field is present (the machine is loaded), the no load mmf on the pole must be made greater than that represented by point *P*; it should be equal to the value corresponding to point *Q*.

The difference between the mmfs represented by points *Q* and *P* in Figure A2.6 represents the demagnetising mmf resulting from the cross-magnetising mmf of the armature reaction field. This is the amount by which the no load pole ampere-turns (AT_{fo}) must be increased so that the design value of the average flux per pole in the air-gap (ϕ) is obtained when the machine is loaded.

APPENDIX 3

THE MAGNETIC CIRCUIT OF INDUCTION MOTORS

A3.1 THE MAGNETISING CURRENT

The magnetic circuit of a 4-pole motor is shown in Figure A3.1. It is seen that the flux produced by each pole of the stator winding divides in the rotor core, one half returning through the rotor teeth, air-gap, stator teeth and stator core of each of the adjacent poles.

The magnetic circuit calculations for an induction motor follows the same general treatment as in d.c. machines. The main difference is that unlike the d.c. machine, where the flux is assumed to be constant and uniformly distributed over any section, in an induction motor the flux is, approximately, sinusoidally distributed.

If we can ignore saturation, the permeability of the various iron sections will be constant and accurate results can be obtained from calculations based on the average value of mmf and the flux tube where this mean occurs. However, saturation

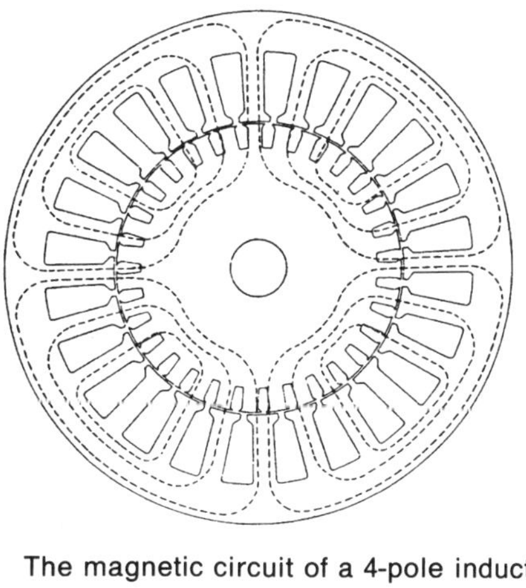

Figure A3.1 The magnetic circuit of a 4-pole induction motor.

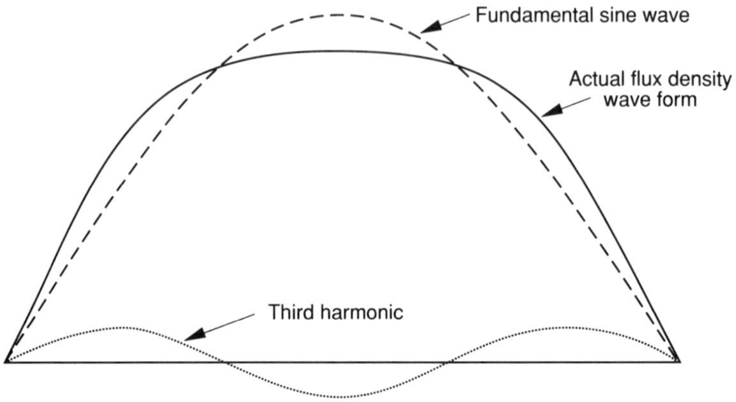

Figure A3.2 Effect of saturation on the shape of air-gap flux density waveform.

invariably occurs, particularly, in the stator and rotor teeth. As a result, the air-gap flux density waveform will have a flat top, as shown in Figure A3.2.

Calculations based on the mean value of the sinusoidally distributed flux yields an under-estimated value of magnetising current. On the other hand, using the maximum flux density value, and assuming a sinusoidal flux density distribution, over-estimates the magnetising current.

The flux density distribution in space can be approximated to a reasonable accuracy with a fundamental sine wave and a third harmonic. The flux density value at an angle of 60 electrical degrees from the interpolar axis is due to the fundamental sine wave alone (the third harmonic vanishes at this point), and this value lies on the *actual* curve. Therefore, calculations of magnetising mmf for the air-gap and teeth are based on the flux density value at 60 electrical degrees from the interpolar axis. The total ampere-turns per pole required to establish the flux in the magnetic circuit of an induction motor is:

$$AT_p = AT_g + AT_{ts} + AT_{tr} + AT_{cs} + AT_{cr} \tag{A3.1}$$

The root-mean-square value of the magnetising current per phase is:

$$I_m = \frac{0.427pAT_p}{T_s K_{ws}} \tag{A3.2}$$

A3.1.1 Air-gap ampere-turns

The ampere-turns per pole required to pass the flux in the air-gap is:

$$AT_g = H_{g60} l_g K_g = \frac{B_{g60}}{\mu_0} l_g K_g = \frac{1.36 B_{av}}{\mu_0} l_g K_g \tag{A3.3}$$

where B_{av} is the average flux density in the air-gap, T, l_g is the radial length of the air-gap, m and K_g is the air-gap correction factor.

As the slot openings of both stator and rotor act to increase the air-gap reluctance (or, effectively, its radial length), the correction factor K_g is the product of the stator and rotor air-gap coefficients.

A3.1.2 Stator teeth ampere-turns

The laminations used for small induction motors have tapered slots and parallel-sided teeth. The tooth flux density is constant across all sections and it is given by:

$$B_{t60} = \frac{1.36\phi}{(S_s/p)L_i W_{ts}} \tag{A3.4}$$

where ϕ is the average flux per pole, Wb, S_s is the number of stator slots, L_i is the core iron length, m and W_t is the tooth width, m.

The factor 1.36 is used in equation (A3.2) to give the flux per pole corresponding to a point 60 electrical degrees from the interpolar axis.

The ampere-turns per metre at_{ts} for the stator teeth is found from the magnetisation characteristics of the lamination material. The ampere-turns required to pass the flux in the stator teeth is then:

$$AT_{ts} = at_{ts} \times d_{ss} \tag{A3.5}$$

where d_{ss} is the depth of stator slot, m.

A3.1.3 Rotor teeth ampere-turns

Using the number of rotor slots and the width of the rotor tooth in equation (A3.4) yields the flux density in the rotor tooth. The ampere-turns per metre at_{tr} for the rotor teeth is found from the magnetisation characteristics of the lamination material. The ampere-turns required to pass the flux in the rotor teeth is:

$$AT_{tr} = at_{tr} \times d_{sr} \tag{A3.6}$$

where d_{sr} is the depth of the rotor slot, m.

A3.1.4 Stator and rotor core ampere-turns

In induction motors, the flux does not enter the stator and rotor cores at a point, as is evident in Figure A3.1. Therefore, the core density is not constant over the length of the flux path. If the air-gap flux density produced by the distributed winding is near sinusoidal, the core flux density will also be sinusoidally distributed in space with a maximum value on the interpolar axis.

The mmf per pole for the stator and rotor cores can be calculated by assuming a maximum value of sinusoidally varying core flux density and determining the ampere-turns per metre at various points of the wave. By repeating such calculations for a number of maximum core flux density values, once and for all, a relationship between B_{cmax} and average at_c can be obtained for a given lamination material.

When the laminations magnetisation characteristics are available in terms of the maximum core density and the average ampere-turns per metre, the length of the flux path in the cores will be taken equal to half the pole pitch on the mean diameter of the core. For the stator:

$$l_{cs} = \frac{\pi(D + 2d_{ss} + d_{cs})}{2p} \tag{A3.7}$$

and for the rotor:

$$l_{cr} = \frac{\pi(D_r - 2d_{sr} - d_c)}{2p} \tag{A3.8}$$

where (A3.9).

$$D_r = D - 2l_g \tag{A3.9}$$

If the maximum core flux density versus the average ampere-turns per metre curve is not available, one may use the average core density corresponding to the average flux per pole. In such a case, the length of the flux path in the core will be taken as 1/3 pole pitch on the mean diameter of the core.

The ampere-turns per pole for the stator core is:

$$AT_{cs} = at_{cs}l_{cs} \tag{A3.10}$$

and for the rotor core:

$$AT_{cr} = at_{cr}l_{cr} \tag{A3.11}$$

A3.2 REACTANCE CALCULATIONS

The winding leakage reactance of an electric machine is calculated by defining the leakage flux paths and calculating their length and area. For induction motors, it is common practice to divide the total leakage flux into: stator and rotor leakage, zigzag leakage, end-connection leakage, belt or differential leakage and skew leakage.

Slot leakage results because not all the flux crossing the air-gap links all of the conductor area in the slot. The slot tapering angle is usually related to the number of slots used, not the lamination dimensions. Slot leakage increases as ratio of the slot depth to width increase.

The permeance coefficient (or specific permeance) of partially closed slots usually encountered in small induction motors may be calculated with the aid of Figure A3.3.

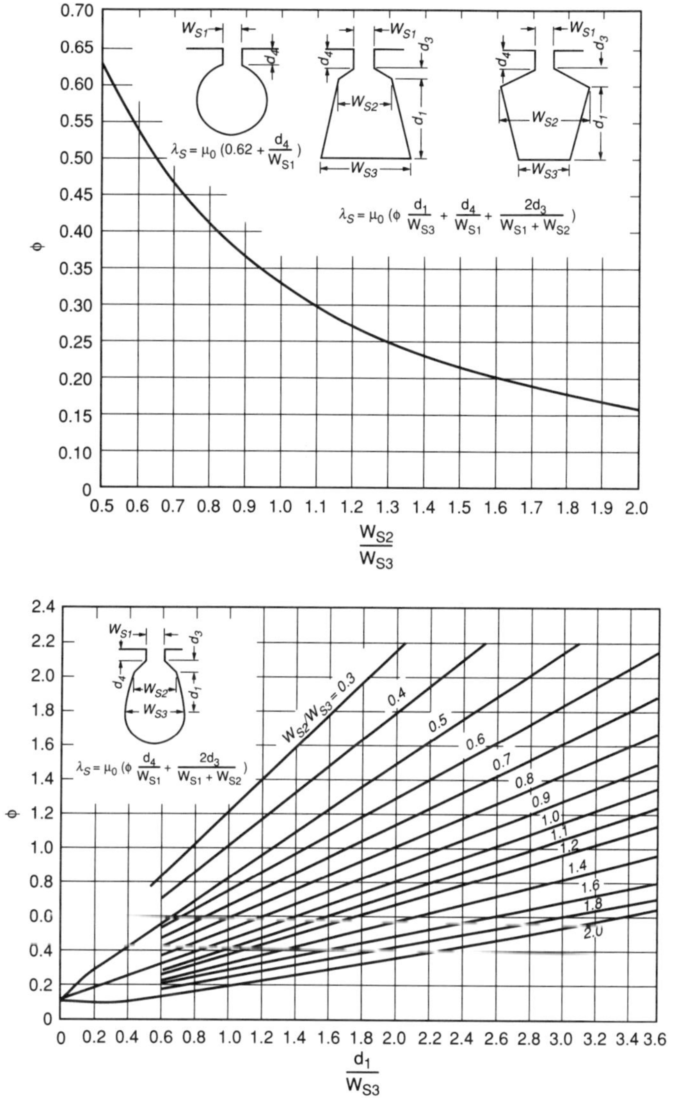

Figure A3.3 The calculation of slot permeance coefficient.

The rotor slot leakage must be referred to stator, therefore:

$$\lambda_s = \lambda_{ss} + \lambda_{sr} \frac{K_{ws}^2 S_s}{K_{wr}^2 S_r}$$

where λ_{ss} and λ_{sr} are the permeance coefficient values for the stator and rotor slots, respectively.

The total slot leakage reactance per phase is:

$$x_s = \frac{8\pi f T_s^2 L}{pq} \lambda_s \ \Omega \tag{A3.12}$$

Zigzag leakage flux passes from one tooth to another on opposite sides of the air-gap in a zigzag fashion. It depends upon the radial air-gap length and the relative positions of teeth tips. Zigzag specific permeance may be calculated from:

$$\lambda_z = \mu_0 \frac{W_{ts} W_{tr} (W_{ts}^2 + W_{tr}^2)}{12 l_g \tau_{ss}^2 \tau_{sr}}$$

where W_{ts} and W_{tr} are the width of the stator and rotor teeth at the air-gap, respectively, m, τ_{ss} and τ_{sr} are the stator and rotor slot pitch, respectively, m and l_g is the radial air-gap length, m.

The zigzag leakage reactance per phase is:

$$x_z = \frac{8\pi f T_s^2 L}{pq} \lambda_z \ \Omega \tag{A3.13}$$

The end connections, or the overhang, leakage flux depends on the arrangements of the end connections, on their proximity to the end covers, cores and frame, and on the magnetic properties of the end covers. Therefore, the exact calculation of this component of leakage flux is extremely difficult, almost impossible. Calculations based on ideal coil shapes yield empirical formulae which provide approximate values. One such formula for the overhang permeance coefficient is:

$$\lambda_0 = \mu_0 \frac{K_o \tau^2}{\pi L_0 \tau_{ss}} \tag{A3.14}$$

where τ is the pole pitch, m, L_0 is the length of conductor in the end connection, m, and K_o is the dimensionless overhang factor which may be taken from Figure A3.4.

The overhang leakage reactance per phase is:

$$x_o = \frac{8\pi f T_s^2 L_0}{pq} \lambda_0 \quad \Omega \tag{A3.15}$$

The belt leakage reactance is negligible in squirrel-cage motors having an integral number of slots per pole. For wound rotor motors, the belt leakage reactance per phase for stator and rotor may be calculated in terms of the magnetising current per phase due to the air-gap ampere-turns (I_{mg}) as follows:

$$x_b = \frac{E}{I_{mg}} (K_{bs} + K_{br}) \quad \Omega \tag{A3.16}$$

where K_{bs} and K_{br} are the belt leakage constants and they may be obtained from Figure A3.5.

When applicable, the skew leakage reactance is obtained in terms of the skew angle θ_{sk} expressed in radians as follows:

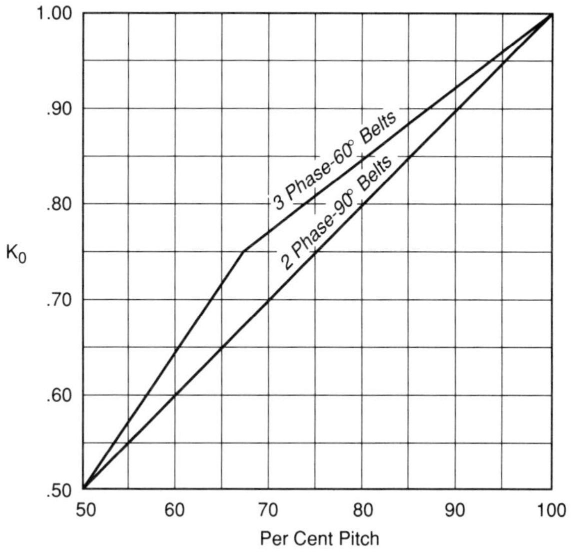

Figure A3.4 Overhang factor.

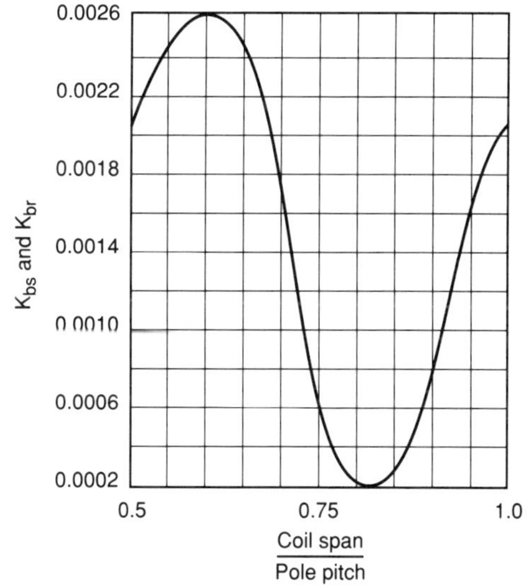

Figure A3.5 Belt-leakage constants.

$$x_{\rm sk} = \frac{E}{I_{\rm mg}} \frac{\theta_{\rm sk}^2}{12} \tag{A3.17}$$

where $I_{\rm mg}$ is the magnetising current per phase due to the air-gap ampere-turns.

APPENDIX 4

COMMUTATION DECODER DESIGN

A4.1 DERIVATION OF SWITCHES BOOLEAN FUNCTIONS

The truth table of the commutation decoder is given in chapter 6 (Table 6.3, page 173) for the case of three sensors. The Boolean function for switch T1 is readily expressed as follows:

$$
\begin{aligned}
T1 &= \overline{F}.H1.\overline{H2}.H3 + \overline{F}.H1.\overline{H2}.\overline{H3} + R.\overline{H1}.H2.H3 + R.\overline{H1}.H2.\overline{H3} \\
&= \overline{F}.H1.\overline{H2}.(H3 + \overline{H3}) + R.\overline{H1}.H2(H3 + \overline{H3}) \qquad \text{(A4.1)}
\end{aligned}
$$

Using De Morgan's laws:

$$A + \overline{A} = 1, \quad \overline{A + B)} = \overline{A}.\overline{B} \qquad \text{(A4.2)}$$

Applying the first law to equation (A4.1) we obtain:

$$T1 = \overline{F}.H1.\overline{H2} + R.\overline{H1}.H2 \qquad \text{(A4.3)}$$

Taking double complement of the RHS of equation (A4.3) and applying the second law of equation (A4.2), we get:

$$T1 = \overline{\overline{(\overline{F}.H1.\overline{H2})}\;\overline{(R.\overline{H1}.H2)}} \qquad \text{(A4.4)}$$

Similarly, the Boolean functions of the other switches can be obtained. They are as follows:

$$T2 = \overline{\overline{(\overline{F}.\overline{H1}.H2)}\;\overline{(R.H1.\overline{H2})}} \qquad \text{(A4.5)}$$

$$T3 = \overline{\overline{(\overline{F}.H2.\overline{H3})}\;\overline{(R.\overline{H2}.H3)}} \qquad \text{(A4.6)}$$

$$T4 = \overline{\overline{(\overline{F} \,.\, \overline{H2} \,.\, H3)}\;\overline{(R \,.\, H2 \,.\, \overline{H3})}} \tag{A4.7}$$

$$T5 = \overline{\overline{(\overline{F} \,.\, H3 \,.\, \overline{H1})}\;\overline{(R \,.\, \overline{H3} \,.\, H1)}} \tag{A4.8}$$

$$T6 = \overline{\overline{(\overline{F} \,.\, \overline{H3} \,.\, H1)}\;\overline{(R \,.\, H3 \,.\, \overline{H1})}} \tag{A4.9}$$

INDEX